Visualizing Linear Models

W. D. Brinda

# Visualizing Linear Models

Springer

W. D. Brinda
Statistics and Data Science
Yale University
New Haven, CT, USA

ISBN 978-3-030-64169-6 ISBN 978-3-030-64167-2 (eBook)
https://doi.org/10.1007/978-3-030-64167-2



This Springer imprint is published by the registered company Springer Nature Switzerland AG
The registered company address is: Gewerbestrasse 11, 6330 Cham, Switzerland

*Dedicated to my parents.*

# About the Book

This book accompanies my one-semester Linear Models (S&DS 312/612) course at Yale University. It provides a visual and intuitive coverage of the core theory of linear models. Designed to also develop fluency with the underlying mathematics and to impart a deep understanding of the principles, it gives graduate students and advanced undergraduates a solid foundation to build upon.

## Structure

Three chapters *gradually* develop the essentials of linear model theory. A chapter on least-squares fitting is provided before any discussion of probability. This helps clarify that the least-squares procedures can be understood without any reference to modeling. A later chapter introduces the linear model and explores properties of least-squares estimators in the context of such models. This chapter works through a variety of results that do not depend on assuming any particular distribution for the errors. The final chapter adds the assumption that the linear model's errors are Normally distributed and develops additional inference techniques that this Normality assumption makes possible.

## Two Types of Pictures

As the title indicates, one of the most important aspects of this book's presentation is its focus on *visualization*. Two distinct and complementary types of pictures are explained. The scatterplot is an example of what I call the *observations picture*; each *observation* (row of a data frame) is a point in the picture. Observations pictures are drawn to help one better understand a particular dataset. On the contrary, the *variables picture* draws each *variable* (column of a data frame) as a vector. Such pictures are not drawn to help you understand a particular dataset; rather, a *generic* version of the variables picture is invaluable for making sense of a great deal of linear model theory. For example, the variables picture makes it clear that least-squares fitting is simply a case of orthogonal projection. In addition, "sum-of-squares" decompositions can be easily understood via the variables picture as instances of the Pythagorean theorem. The variables pictures are challenging to comprehend at first, but the book's gradual approach helps. By starting with only the data, the text introduces a bare-bones version of the picture that gently eases the reader into this powerful new way of visualizing data. Several of the book's key figures are recollected and juxtaposed for easy comparison in an appendix section.

## Background Topics

This book has three chapters on linear model theory, each of which is preceded by a concise and targeted *background* chapter covering essential mathematical concepts and tools. The background chapters are designed to help the reader develop a high level of fluency with the *ingredients* of linear model theory. Because they are also the ingredients of a broad spectrum of other quantitative fields and applications, this part of the book is every bit as valuable as its coverage of linear models.

The background chapters should not be considered a comprehensive survey of their subject matter; they omit topics that are not directly used in the book's development of linear model theory. For instance, the linear algebra chapter does not bother introducing *determinants* because the linear model theory chapters do not make use of them.

Some aspects of the background topics are covered at a high level of abstraction, which could make them challenging for a beginner. However, my goal is for the abstraction to simplify and unify the material for an intermediate student. How could abstract coverage possibly be simpler? It forces the reader to focus on the essentials. Most of the important properties of vectors in $\mathbb{R}^2$ do not have anything to do with being able to represent them as an ordered pair of real numbers. Rather the most useful properties are more fundamental to the nature of how the vectors can be added and subtracted and how they can multiply and divide real numbers. Obviously, abstraction can allow you to represent your knowledge more efficiently. If several important results are all simple consequences of a more abstract statement, they do not necessarily have to be stored separately in your brain. Finally, abstraction can also make exercises and theorems slightly less wordy. For example, I prefer "show that the span of $\mathbf{v}_1, \ldots, \mathbf{v}_m$ is a subspace" to "show that the span of $\mathbf{v}_1, \ldots, \mathbf{v}_m \in \mathbb{R}^n$ is a subspace" when there is absolutely no need to introduce another symbol $n$ or to indicate a way of explicitly representing the vectors.

If you find the abstract definitions difficult to wrap your head around, that should not be an obstacle to comprehending these background chapters in the main. You are encouraged to treat basic facts about these abstract constructions as axiomatic, though their derivations are available in the Solutions for the interested reader. Facts established in abstract settings (e.g., for *vectors* or *random vectors*) can be freely applied to more concrete instantiations (e.g., *vectors in* $\mathbb{R}^n$ or $\mathbb{R}^n$*-valued random vectors*).

## Chunking and Purposeful Practice

Imagine that you want to become as good as possible at tennis. How should you spend your time? One option is to read book after book about playing tennis. You will learn some useful things, but reading by itself will do very little for your tennis game. Alternatively, you could find a partner and spend all of your time actually playing tennis matches. This will surely make you better, but it will not come close to making you the best player you can be. You will plateau pretty quickly. A combination of reading and playing would be better than either one by itself, but ultimately you are still missing what is by far the most important activity for improving performance in tennis or any other sport: *practice exercises*. Tennis performance involves numerous *ingredients*, such as forehand, backhand, serving, and so on; those skills can be further broken down into component parts. For every little component, you can find practice exercises designed to perfect your execution of that action. You will do the exact same exercise over and over rapidly, trying to notice and correct imperfections, ideally with the assistance of an expert coach.

Importantly, you do not *stop* practicing a technique once you have gotten it right. Understanding how to do it correctly is the first step, but understanding is not enough. On the contrary, when you have finally gotten a technique right, that is when you are truly able to *start* practicing it.

Here is the key: these drills are literally building neural circuitry in your brain that will later enable you to execute the action correctly *automatically*, that is, *using as little "RAM" as possible*. During a real match, your RAM will instead be dedicated to reacting and to high-level strategizing.

This is not just about sports, of course, or I would not have bothered discussing it. Performance in intellectual domains works the same way. According to renowned psychologist K. Anders Ericsson who pioneered the study of expert performance,

> [Experts'] years of practice have changed the neural circuitry in their brains to produce highly specialized mental representations... The key benefit of mental representations lies in how they help us deal with information: understanding and interpreting it, holding it in memory, organizing it, analyzing it, and making decisions with it... [These] make possible the incredible memory, pattern recognition, problem solving, and other sorts of advanced abilities needed to excel in their particular specialities. (Ericsson and Pool 2017)

The mind can only deal with a few *items* at a time. Experts overcome this limitation by "chunking" multiple ideas together into a single *item*, allowing them to be held in the mind with more RAM to spare. Over time, complex items continue to combine with each other, and eventually the mind is capable of processing ideas that would have been literally unthinkable without all that fancy new circuitry. Neurologically speaking, chunking means developing specialized neural circuitry not unlike what is needed for an athlete to automate certain actions. *The most effective way to build this circuitry is with exercises* resembling those that are routine to athletes and musicians.

So, what are the characteristics of exercises that build this neural circuitry effectively? Here are four principles of what Ericsson calls *purposeful practice*:

- *Well-defined task*: Be specific about what you are trying to do, for instance, solve a problem, explain a concept from recall, etc.
- *Appropriate level of difficulty for the individual*: Purposeful practice tasks should feel like a stretch. If they are too easy, then you are not improving much and/or you might become bored. If they are too hard, then you are not getting anywhere and you might become demoralized. Lifting weights is a good metaphor.
- *Rapid and informative feedback*: There should be some way for you to tell as rapidly as possible how well you are doing. For example, you may have a coach present, or you may be working through a problem with a known solution that you can refer to when needed.
- *Opportunities for repetition and correction of errors*: Use the feedback to adjust your technique. You will need to concentrate hard on what you are doing to accomplish this as effectively as possible. And, once you are doing something perfectly, try to do it perfectly over and over to commit the process to memory. Because of the intense concentration required, you should feel mentally tired after a purposeful practice session.

*Engaging in purposeful practice exercises appears to be the most efficient way to build the neural circuitry that enables expert performance.*

The usual academic tasks do not conform to these principles well, especially with regard to rapid feedback and opportunities for repetition. Think about math homework, for example. The problems are often too challenging and therefore inordinately time consuming, feedback comes much later (if ever), and naturally the problems are not repeated. Of course, people do *eventually* develop expertise through these activities, but based on my understanding of the science of expert performance, it is not a very efficient approach.

Let us think about an alternative way that you might spend your time to develop expertise in a mathematical domain. In mathematics, or any intellectual domain, we can be more specific about what an effective purposeful practice exercise should entail. Embrace one of the most robust results in the study of learning: *recall tasks* are far more effective than other common reviewing techniques for building knowledge into your brain (Brown et al. 2014). A recall task is an activity in which you are presented with a prompt and you then have to state or write a specific correct response that you have learned before.[1] Think *flashcards*. Indeed, *quizzing yourself with well-designed flashcards is a terrific purposeful practice activity; it is highly effective at building the neural circuitry that enables expert performance*.

You may protest that flashcards are fine for rote memorization of anatomy or of words in a foreign language, but that is different from expertise at *doing mathematics*. This difference is not as consequential as you might think. In my experience, flash cards are ideal for math as well, they just need to be designed properly. To be effective at mathematics, you need to understand a large body of *facts*, and you also need to be able to reproduce a hefty collection of *patterns of*

[1]The correct answer should not be embedded in the prompt, as in a multiple choice question.

*reasoning*. An ideal mathematics flashcard will reinforce both. It will require a short derivation that exemplifies a powerful pattern of reasoning *and* proves a widely useful fact. Not every card has to fit this schema, but it is good to keep the ideal in mind.

But what about creativity? If you are just reinforcing things you have already read, you will not practice having to generate your own proofs and answers. Again, I will speak from my own experience. Several years into my doctoral degree, I switched from spending my time trying to answer research questions to spending my time trying to develop expertise (that is, build neural circuitry via purposeful practice) most relevant to my research questions. I cannot emphasize enough how much that change improved my progress. I found that once I have developed the neural circuitry to hold a complex idea in my head with RAM to spare, creativity is spontaneous. That spare RAM naturally tries to extend the idea, connect it to other ideas, apply it in new ways, and so on.[2] Usually, my new ideas happen during a practice session when I am working through a flash card for the tenth or twentieth time.

It takes a great deal of time and effort for a student to translate material into high-quality flash cards. To save you the time, I have built the question-and-answer format into this book's presentation of the material. The book itself thus provides a bank of ready-made exercise prompts with which to quiz yourself. These exercises are available on my website (http://quantitations.com) as notecards for Anki, a flashcard app that automatically incorporates the principles of optimal *spaced repetition* of questions.[3] There is also a printable version of the cards for readers who prefer a hard copy.

The prompts in this book are color coded by importance. The *red* questions are the most crucial, while the *blue* questions are not as central; the answers to red and blue questions are presented as they arise in the body of the text. There are also a number of *green* questions, especially in the early parts of the *background* chapters, which should be considered low priority. Many of these are included only to make the book more self-contained. You should read these questions and try to internalize any results they state, but you do not need to worry about their derivations (which can be found in the Solutions at the end of the book).

I recommend setting aside a specific block of time (between 30 min and an hour) every day for a flashcard session; schedule it for the early morning if you are a morning person or for the early evening otherwise. During the allotted period, focus completely on the practice. Make sure to turn off all notifications; try your best to prevent any possible interruption no matter how minor. As an important bonus, these sessions will help rebuild your capacity to *focus* for long spans of time, an

[2]One could argue that I would not be "spontaneously creative" without all those years of math homework problems. I do not know. My gut feeling is that creative insights are much more attributable to having spare RAM than to having practiced creativity. As such, I recommend putting the overwhelming majority of your effort into doing flashcards, though working out unfamiliar problems has its place.

[3]For a thorough exposition of spaced repetition, see http://gwern.net/Spaced-repetition.

ability that has severely deteriorated with the rise of the internet and particularly smartphones.[4]

Here is another good principle to keep in mind for these sessions: focus on the *process* rather than the *product* (Sterner 2012). In the preceding discussion, I have been acting as if there is some ultimate goal of "expertise" that you will eventually reach. If you decide that you are not going to be satisfied until you are an "expert," you will probably never be satisfied. In reality, there is no endpoint; there is always room to continue growing. You will always wish that you had more knowledge than you do. Instead, *take satisfaction in your daily growth* especially since you are engaging in a practice that builds your skills about as effectively as possible. The mindset is not just about your psychological well-being; it also improves the quality of your practice by keeping you focused on the present. Learning does not have to be stressful or burdensome. On the contrary, a daily purposeful practice routine can be grounding and deeply enjoyable.

## For Instructors

It might feel strange to teach a math class without requiring the customary demanding problem sets. How do you still make sure your students are putting time and effort into the class week after week? And, how do you assign grades without creating a small number of stressful high-stakes exams? Here is my approach.

For the last 10 min of every class meeting, a quiz is administered. It asks two questions that are randomly chosen from the cumulative bank of *red* prompts that have been covered so far.

As extra encouragement to use Anki, my students can submit their practice statistics each weekend, exported from the Anki program. Any student who did an adequate number of recall sessions in a given week will get an automatic perfect score on one of their quizzes the following week (replacing a less-than-perfect score from that week if needed).

In place of lengthy problem sets, I assign short computational tasks that require my students to implement in R code something they are currently learning. I aim for this homework to take only about an hour or two per week. Past homework assignments with solutions are available at the book's website.

As a final note, the *background* chapters are not fully covered in the lectures, especially Chap. 1 for which I only discuss orthogonal projection, spectral decomposition, quadratic forms, and principal components, all in the context of $\mathbb{R}^n$ rather introducing the book's extra level of abstraction. Students can read these chapters to differing extents depending on their needs and levels of interest.

[4]For more information on restoring your ability to focus, I will refer the reader to Goleman (2013), Newport (2016), and Carr (2010).

## Sequel

For the serious student of probability or mathematical statistics, an analogous two-pictures approach to understanding random variables and random vectors will be explained in exquisite detail in a follow-up book *Visualizing Random Variables*.

## Acknowledgments

I have been blessed with too many wonderful teachers and peers to list. I am most indebted to Professor Joseph Chang who taught me linear model theory long ago and has for years encouraged me to continue writing. I am also thankful to my students for their valuable feedback and for catching my mistakes. Most importantly, I thank my family for their continual support.

# Contents

# Chapter 1
# Background: Linear Algebra

We will begin our course of study by reviewing the most relevant definitions and concepts of linear algebra. We will also expand on various aspects of orthogonal projection and spectral decomposition that are not necessarily covered in a first linear algebra course. This chapter is almost entirely self-contained as it builds from the ground up everything that we will need later in the text. Two exceptions are Theorems 1.3 and 1.4 which point to external references rather than being proven in this book or its exercise solutions.

Additionally, this chapter's coverage will encourage you to think more abstractly about vector spaces. A linear algebra course focuses on Euclidean space, but many important results also hold *without any additional complications* for other important vector spaces including spaces of random variables.[1] Eventually, we will cover topics for which the abstract theory is decidedly more complicated, so at that point our focus will shift to $\mathbb{R}^n$.

## 1.1 Vector Spaces

A concise definition of vector spaces is provided here, enabling us to state results throughout this chapter in great generality when possible. The main point is to convey that vector spaces in general work a lot like familiar Euclidean spaces. If you would like to see more formal (and more parsimonious) definitions, consult an abstract algebra text such as Artin (2010), Ch 3. On the other hand, if you are overwhelmed by the abstract nature of this section or the next, you need only glance over them before moving on to Sect. 1.3; throughout the remainder of this chapter, any "vector" in a "vector space" can essentially be treated as a vector in Euclidean

[1] Although the applications in this text only require Euclidean space, the extra abstraction from this chapter will be put to good use in a sequel volume.



W. D. Brinda, *Visualizing Linear Models*,
https://doi.org/10.1007/978-3-030-64167-2_1

space and any "scalar" can essentially be treated as a real number. Or you can skip ahead to Chap. 2 and come back to this chapter for reference as needed.

**Definition (Field)** A **field** is a set of elements that can be added and multiplied together to produce other elements of the set.

- *operator properties*: Addition and multiplication are both associative and commutative, and multiplication distributes over addition.
- *identity elements*: The field contains unique elements 0 and 1, respectively, satisfying $a + 0 = a$ and $a \cdot 1 = a$ for every $a$ in the field.
- *inverse elements*: For every $a$ in the field, there exists an element $-a$ such that $a+(-a) = 0$, and if $a \neq 0$, there also exists an element $a^{-1}$ such that $a \cdot a^{-1} = 1$.

As usual, *subtraction* and *division* are short-hand notation for certain additions and multiplications. $a - b$ means $a + (-b)$ and $a/b$ means $a \cdot b^{-1}$. Familiar fields include the rational numbers, the real numbers, and the complex numbers.

**Definition (Vector Space)** A **vector space** is a set of objects called **vectors** that can be added together to produce other elements of the set, along with a field whose elements (called the **scalars** of the vector space) can multiply the vectors.

- *operator properties*: Vector addition is associative and commutative, scalar multiplication is associative (whether multiplying vectors or other scalars), and scalar multiplication distributes over vector addition and over addition of scalars.
- *identity elements*: There is a unique vector $\mathbf{0}$ satisfying $\mathbf{v}+\mathbf{0} = \mathbf{v}$ for every vector $\mathbf{v}$ in the vector space.
- *inverse elements*: For every vector $\mathbf{v}$, there exists a vector $-\mathbf{v}$ such that $\mathbf{v}+(-\mathbf{v}) = \mathbf{0}$.

In particular, any vector space whose scalar field is the real numbers is called a **real vector space**.

Note that each vector in the vector space $\mathbb{R}^n$ is represented as an ordered list of $n$ real numbers and the reals are also taken to be the scalar field.

**Exercise 1.1** Let $\mathbf{0}$ be the zero vector, and let $a$ be a scalar. Show that $a\mathbf{0} = \mathbf{0}$.

**Exercise 1.2** Let $\mathbf{v}$ be a vector, and let 0 be the zero scalar. Show that $0\mathbf{v}$ equals the zero vector.

## 1.2 Linear Combinations

**Definition (Linear Combination)** Let $\mathbf{v}_1, \ldots, \mathbf{v}_m$ be vectors in a vector space. For any scalars $b_1, \ldots, b_m$, the sum $b_1\mathbf{v}_1 + \ldots + b_m\mathbf{v}_m$ is called a **linear combination** of the vectors.

Let us write the linear combination $b_1\mathbf{v}_1 + \ldots + b_m\mathbf{v}_m$ with a notation that resembles matrix-vector multiplication; think of the vectors as columns of a matrix and the scalars as entries of a vector multiplying that matrix on the right.

$$\begin{bmatrix} \mathbf{v}_1 & \cdots & \mathbf{v}_m \end{bmatrix} \begin{bmatrix} b_1 \\ \vdots \\ b_m \end{bmatrix} := b_1\mathbf{v}_1 + \ldots + b_m\mathbf{v}_m \tag{1.1}$$

The set of all possible coefficients of the vectors $\mathbf{v}_1, \ldots, \mathbf{v}_m$ is itself a vector space (denoted $\mathcal{F}^m$ where $\mathcal{F}$ is the scalar field). Letting[2] $\mathbf{b} := (b_1, \ldots, b_m)$ and $\mathbf{c} := (c_1, \ldots, c_m)$ be "vectors of scalars" and letting $a_1$ and $a_2$ be scalars, we can take the linear combination to be

$$\begin{aligned} a_1\mathbf{b} + a_2\mathbf{c} &= a_1 \begin{bmatrix} b_1 \\ \vdots \\ b_m \end{bmatrix} + a_2 \begin{bmatrix} c_1 \\ \vdots \\ c_m \end{bmatrix} \\ &= \begin{bmatrix} a_1 b_1 \\ \vdots \\ a_1 b_m \end{bmatrix} + \begin{bmatrix} a_2 c_1 \\ \vdots \\ a_2 c_m \end{bmatrix} \\ &= \begin{bmatrix} a_1 b_1 + a_2 c_1 \\ \vdots \\ a_1 b_m + a_2 c_m \end{bmatrix} \end{aligned}$$

which is yet another vector of scalars.

We can verify that the linear combination notation distributes in the ways that matrix-vector multiplication does. By associativity and commutativity,

$$\begin{aligned} \begin{bmatrix} \mathbf{v}_1 & \cdots & \mathbf{v}_m \end{bmatrix}(\mathbf{b} + \mathbf{c}) &= \begin{bmatrix} \mathbf{v}_1 & \cdots & \mathbf{v}_m \end{bmatrix} \begin{bmatrix} b_1 + c_1 \\ \vdots \\ b_m + c_m \end{bmatrix} \\ &= (b_1 + c_1)\mathbf{v}_1 + \ldots + (b_m + c_m)\mathbf{v}_m \\ &= (b_1\mathbf{v}_1 + \ldots + b_m\mathbf{v}_m) + (c_1\mathbf{v}_1 + \ldots + c_m\mathbf{v}_m) \\ &= \begin{bmatrix} \mathbf{v}_1 & \cdots & \mathbf{v}_m \end{bmatrix}\mathbf{b} + \begin{bmatrix} \mathbf{v}_1 & \cdots & \mathbf{v}_m \end{bmatrix}\mathbf{c}. \end{aligned} \tag{1.2}$$

We can also factor scalars out of the linear combination notation. For example, with a scalar $a$, the $a\mathbf{b}$ linear combination of $\mathbf{v}_1, \ldots, \mathbf{v}_m$ is the same as $a$ times the $\mathbf{b}$ linear combination:

[2] Although vectors are written out horizontally in this book's paragraph text, they should always be interpreted as *column* vectors in any mathematical expression.

$$
\begin{aligned}
\begin{bmatrix}\mathbf{v}_1 \cdots \mathbf{v}_m\end{bmatrix}(a\mathbf{b}) = \begin{bmatrix}\mathbf{v}_1 \cdots \mathbf{v}_m\end{bmatrix}\begin{bmatrix}ab_1\\ \vdots \\ ab_m\end{bmatrix} \\
&= ab_1\mathbf{v}_1 + \ldots + ab_m\mathbf{v}_m \\
&= a(b_1\mathbf{v}_1 + \ldots + b_m\mathbf{v}_m) \\
&= a\begin{bmatrix}\mathbf{v}_1 \cdots \mathbf{v}_m\end{bmatrix}\mathbf{b}. \qquad (1.3)
\end{aligned}
$$

Additionally, analogous to matrix addition, we can write $\begin{bmatrix}\mathbf{v}_1 \cdots \mathbf{v}_m\end{bmatrix} + \begin{bmatrix}\mathbf{w}_1 \cdots \mathbf{w}_m\end{bmatrix}$ to naturally represent $\begin{bmatrix}\mathbf{v}_1 + \mathbf{w}_1 \ldots \mathbf{v}_m + \mathbf{w}_m\end{bmatrix}$. Then multiplication by a vector of scalars distributes over the sum:

$$
\begin{aligned}
\left(\begin{bmatrix}\mathbf{v}_1 \cdots \mathbf{v}_m\end{bmatrix} + \begin{bmatrix}\mathbf{w}_1 \cdots \mathbf{w}_m\end{bmatrix}\right)\mathbf{b} &= \begin{bmatrix}\mathbf{v}_1 + \mathbf{w}_1 \ldots \mathbf{v}_m + \mathbf{w}_m\end{bmatrix}\mathbf{b} \\
&= b_1(\mathbf{v}_1 + \mathbf{w}_1) + \ldots + b_m(\mathbf{v}_m + \mathbf{w}_m) \\
&= (b_1\mathbf{v}_1 + \ldots + b_m\mathbf{v}_m) + (b_1\mathbf{w}_1 + \ldots + b_m\mathbf{w}_m) \\
&= \begin{bmatrix}\mathbf{v}_1 \cdots \mathbf{v}_m\end{bmatrix}\mathbf{b} + \begin{bmatrix}\mathbf{w}_1 \cdots \mathbf{w}_m\end{bmatrix}\mathbf{b}.
\end{aligned}
$$

The notation introduced and developed in this subsection will be feature prominently in this chapter's early exercises. It may already be familiar based on your experience multiplying matrices with vectors, but now you are justified in using it more generally.

## 1.3 Subspaces

**Definition (Subspace)** Let $\mathcal{S}$ be a set of vectors in a vector space $\mathcal{V}$. If for every pair $\mathbf{v}_1, \mathbf{v}_2 \in \mathcal{S}$ and every pair of scalars $b_1, b_2$, the linear combination $b_1\mathbf{v}_1 + b_2\mathbf{v}_2$ is also in $\mathcal{S}$, then $\mathcal{S}$ is called a **subspace** (of $\mathcal{V}$). Additionally, if there exist vectors in $\mathcal{V}$ that are not also in $\mathcal{S}$, then $\mathcal{S}$ is a **proper subspace** of $\mathcal{V}$.

Note that if $\mathcal{S} \subseteq \mathcal{V}$ is a subspace, then it is also a vector space in its own right.

**Definition (Span)** The **span** of $\mathbf{v}_1, \ldots, \mathbf{v}_m$ is the set of all of their possible linear combinations.

Figure 1.1 shows the span of two vectors in $\mathbb{R}^3$, but we will see that the picture extends to more general settings. Pictures like this will feature prominently in the upcoming chapters.

**Exercise 1.3** Show that the span of $\mathbf{v}_1, \ldots, \mathbf{v}_m$ is a subspace.

*Solution* Consider two vectors in the span, say $\begin{bmatrix}\mathbf{v}_1 \cdots \mathbf{v}_m\end{bmatrix}\mathbf{b}_1$ and $\begin{bmatrix}\mathbf{v}_1 \cdots \mathbf{v}_m\end{bmatrix}\mathbf{b}_2$. For a pair of scalars $a_1, a_2$ the linear combination

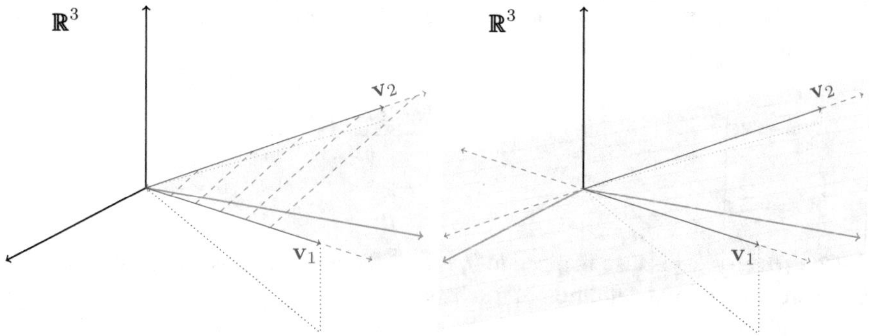


**Fig. 1.1** **Left.** The blue shaded region (and its continuation) represents the linear combination of $\mathbf{v}_1$ and $\mathbf{v}_2$ with non-negative coefficients. It might be helpful to inspect this picture first in order to more easily interpret the next one. **Right.** The span of $\mathbf{v}_1$ and $\mathbf{v}_2$ is a plane in $\mathbb{R}^3$ that includes the origin

$$a_1\begin{bmatrix}\mathbf{v}_1 & \cdots & \mathbf{v}_m\end{bmatrix}\mathbf{b}_1 + a_2\begin{bmatrix}\mathbf{v}_1 & \cdots & \mathbf{v}_m\end{bmatrix}\mathbf{b}_2 = \begin{bmatrix}\mathbf{v}_1 & \cdots & \mathbf{v}_m\end{bmatrix}(a_1\mathbf{b}_1 + a_2\mathbf{b}_2)$$

is also in the span, so the span satisfies the definition of a subspace. ♦

Exercise 1.4 Show that the span of $\mathbf{v}_1, \ldots, \mathbf{v}_m$ is the same as the span of $\mathbf{v}_1 + a_1\mathbf{v}_m, \ldots, \mathbf{v}_{m-1} + a_{m-1}\mathbf{v}_m, \mathbf{v}_m$ for any scalars $a_1, \ldots, a_{m-1}$.

## 1.4 Linear Operators

**Definition** **(Linear Operator)** Let $\mathbb{T}$ be a function that maps from a vector space to another that shares the same scalar field. $T$ is a **linear operator** if $\mathbb{T}(a_1\mathbf{b}_1 + a_2\mathbf{b}_2) = a_1\mathbb{T}(\mathbf{b}_1) + a_2\mathbb{T}(\mathbf{b}_2)$ for all scalars $a_1, a_2$ and vectors $\mathbf{b}_1, \mathbf{b}_2$.

In general, we will omit the parentheses when applying a linear operator, writing, for instance, $\mathbb{T}\mathbf{b}$ rather than $\mathbb{T}(\mathbf{b})$.

Notice that Eqs. 1.2 and 1.3 together show that $\begin{bmatrix}\mathbf{v}_1 & \cdots & \mathbf{v}_m\end{bmatrix}$ is a linear operator. In fact, the following exercise points out that *every* linear operator on $\mathcal{F}^m$ takes this form.

Exercise 1.5 Let $\mathcal{F}$ be a field, and suppose $\mathbb{T}$ is a linear operator from $\mathcal{F}^m$ to $\mathcal{V}$. Show that $\mathbb{T}$ can be represented as a mapping of $\mathbf{b} \in \mathcal{F}^m$ to $\begin{bmatrix}\mathbf{v}_1 & \cdots & \mathbf{v}_m\end{bmatrix}\mathbf{b}$ for some $\mathbf{v}_1, \ldots, \mathbf{v}_m$.

**Definition** **(Null Space)** Let $\mathbb{T}$ be a linear operator. A vector $\mathbf{b}$ is in the **null space** of $\mathbb{T}$ if $\mathbb{T}\mathbf{b} = \mathbf{0}$.

Exercise 1.6 Prove that the null space of $\mathbb{T}$ is a subspace.

*Solution* Let $\mathbf{b}_1$ and $\mathbf{b}_2$ be in the null space. Given any scalars $a_1, a_2$, consider the vector of scalars $a_1\mathbf{b}_1 + a_2\mathbf{b}_2$.

$$\begin{aligned}\mathbb{T}(a_1\mathbf{b}_1 + a_2\mathbf{b}_2) &= a_1 \underbrace{\mathbb{T}\mathbf{b}_1}_{\mathbf{0}} + a_2 \underbrace{\mathbb{T}\mathbf{b}_2}_{\mathbf{0}} \\ &= \mathbf{0}\end{aligned}$$

Since $a_1\mathbf{b}_1 + a_2\mathbf{b}_2$ is also mapped to $\mathbf{0}$, it is in the null space as well; the null space therefore satisfies the definition of a subspace. ♦

**Definition (Affine Operator)** If $\mathbb{T}$ is a linear operator and $\mathbf{v}$ is a vector, then a function $f$ defined by $f(\mathbf{b}) = \mathbf{v} + \mathbb{T}\mathbf{b}$ is called an **affine operator**.[3]

## 1.5 Bases

**Definition (Linear Independence)** Let $\mathbf{v}_1, \ldots, \mathbf{v}_m$ be vectors in a vector space. If no one of these vectors is in the span of the others, then the vectors are said to be **linearly independent**.

Exercise 1.7 Show that if $\mathbf{v}_1, \ldots, \mathbf{v}_m$ are linearly independent, then $(b_1, \ldots, b_m) = (0, \ldots, 0)$ is the only vector of scalars for which $b_1\mathbf{v}_1 + \ldots + b_m\mathbf{v}_m = \mathbf{0}$.

Exercise 1.8 Show that if $(b_1, \ldots, b_m) = (0, \ldots, 0)$ is the only vector of scalars for which $b_1\mathbf{v}_1 + \ldots + b_m\mathbf{v}_m = \mathbf{0}$, then $\mathbf{v}_1, \ldots, \mathbf{v}_m$ are linearly independent.

Exercises 1.7 and 1.8 taken together show that vectors are linearly independent if and only if the zero vector is the only vector in the null space.

Exercise 1.9 Let $\mathbf{z}$ be in the null space of $\begin{bmatrix}\mathbf{v}_1 & \cdots & \mathbf{v}_m\end{bmatrix}$. Given any vector of scalars $\mathbf{b}$, show that the linear combination of $\mathbf{v}_1, \ldots, \mathbf{v}_m$ produced by the entries of $\mathbf{b} + \mathbf{z}$ is exactly the same as that produced by $\mathbf{b}$.

*Solution* With $\mathbf{z}$ in the null space, the $\mathbf{b} + \mathbf{z}$ linear combination results in

$$\begin{aligned}\begin{bmatrix}\mathbf{v}_1 & \cdots & \mathbf{v}_m\end{bmatrix}(\mathbf{b} + \mathbf{z}) &= \begin{bmatrix}\mathbf{v}_1 & \cdots & \mathbf{v}_m\end{bmatrix}\mathbf{b} + \underbrace{\begin{bmatrix}\mathbf{v}_1 & \cdots & \mathbf{v}_m\end{bmatrix}\mathbf{z}}_{\mathbf{0}} \\ &= \begin{bmatrix}\mathbf{v}_1 & \cdots & \mathbf{v}_m\end{bmatrix}\mathbf{b}.\end{aligned}$$

♦

[3] Notice that every *linear* operator is also an *affine* operator: take $\mathbf{v} = \mathbf{0}$ in the definition.

**Exercise 1.10** Show that if $\begin{bmatrix}\mathbf{v}_1 & \cdots & \mathbf{v}_m\end{bmatrix}\mathbf{b} = \begin{bmatrix}\mathbf{v}_1 & \cdots & \mathbf{v}_m\end{bmatrix}\mathbf{c}$, then $\mathbf{c}$ must equal $\mathbf{b} + \mathbf{z}$ for some $\mathbf{z}$ in the null space of $\begin{bmatrix}\mathbf{v}_1 & \cdots & \mathbf{v}_m\end{bmatrix}$.

Exercises 1.9 and 1.10 have together shown that the set of scalar coefficients producing $\begin{bmatrix}\mathbf{v}_1 & \cdots & \mathbf{v}_m\end{bmatrix}\mathbf{b}$ is exactly $\{\mathbf{b} + \mathbf{z} : \mathbf{z} \in \mathcal{N}\}$ where $\mathcal{N}$ denotes the null space of $\begin{bmatrix}\mathbf{v}_1 & \cdots & \mathbf{v}_m\end{bmatrix}$.

**Exercise 1.11** Show that $\mathbf{v}_1, \ldots, \mathbf{v}_m$ are linearly independent if and only if $\mathbf{b} \neq \mathbf{c}$ implies $\begin{bmatrix}\mathbf{v}_1 & \cdots & \mathbf{v}_m\end{bmatrix}\mathbf{b} \neq \begin{bmatrix}\mathbf{v}_1 & \cdots & \mathbf{v}_m\end{bmatrix}\mathbf{c}$, that is, every vector in the span corresponds to a *unique* vector of scalar coefficients.

Exercises 1.7, 1.8, and 1.11 together establish the equivalence in the following definition.

**Definition (Basis)** Let $\mathbf{v}_1, \ldots, \mathbf{v}_m$ be vectors whose span is $\mathcal{V}$. The set $\{\mathbf{v}_1, \ldots, \mathbf{v}_m\}$ is called a **basis** for $\mathcal{V}$ if any (and therefore all) of the following equivalent conditions are true.

- They are linearly independent.
- $(0, \ldots, 0)$ is the only vector in their null space.
- Every $\mathbf{w} \in \mathcal{V}$ has a *unique* vector of scalars $(b_1, \ldots, b_m)$ for which $b_1\mathbf{v}_1 + \ldots + b_m\mathbf{v}_m = \mathbf{w}$.

**Exercise 1.12** Let $\{\mathbf{v}_1, \ldots, \mathbf{v}_m\}$ be a basis for $\mathcal{V}$. How do you know that $\{\mathbf{v}_2, \ldots, \mathbf{v}_m\}$ is not also a basis for $\mathcal{V}$?

**Exercise 1.13** Let $\{\mathbf{v}_1, \ldots, \mathbf{v}_m\}$ be a basis for $\mathcal{V}$, and let $\mathcal{S}$ be a *proper* subspace of $\mathcal{V}$. Explain why at least one of $\mathbf{v}_1, \ldots, \mathbf{v}_m$ is not in $\mathcal{S}$.

**Exercise 1.14** Suppose $\{\mathbf{v}_1, \ldots, \mathbf{v}_m\}$ is a basis for $\mathcal{V}$. Prove that *every* basis for $\mathcal{V}$ has exactly $m$ vectors.

Exercise 1.14 provides a rationale for the following definition (at least for vector spaces that have a basis with finitely many vectors).

**Definition (Dimension)** Let $B \subseteq \mathcal{V}$ be a basis for $\mathcal{V}$. The number of vectors in $B$ is called the **dimension** of $\mathcal{V}$. The dimension of the *trivial* subspace $\{\mathbf{0}\}$ is defined to be 0.

**Exercise 1.15** Let $\mathcal{V}$ be an $m$-dimensional vector space. Prove that any set of $m$ linearly independent vectors in $\mathcal{V}$ must be basis for $\mathcal{V}$.

**Exercise 1.16** Let $\mathcal{S}$ be a subspace of an $n$-dimensional vector space $\mathcal{V}$. Prove that a basis for $\mathcal{S}$ exists.

**Exercise 1.17** Let $\mathcal{F}$ be a field. Find the dimension of $\mathcal{F}^m$ as defined in Sect. 1.2.

*Solution* Consider the $n$ vectors $\mathbf{e}_1 := (1, 0, \ldots, 0), \ldots, \mathbf{e}_m := (0, \ldots, 0, 1)$. A given vector $(c_1, \ldots, c_m) \in \mathcal{F}^m$ has the unique representation $c_1\mathbf{e}_1 + \ldots + c_m\mathbf{e}_m$ with respect to these vectors, so they comprise a basis (known as the *standard basis*). This tells us that the dimension of $\mathcal{F}^m$ is $m$. ♦

**Exercise 1.18** Suppose $\begin{bmatrix}\mathbf{v}_1 \cdots \mathbf{v}_m\end{bmatrix}$ and $\begin{bmatrix}\mathbf{w}_1 \cdots \mathbf{w}_m\end{bmatrix}$ have the exact same behavior on a basis $B = \{\mathbf{b}_1, \ldots, \mathbf{b}_m\}$ for the vector space of scalar coefficients, that is, $\begin{bmatrix}\mathbf{v}_1 \cdots \mathbf{v}_m\end{bmatrix}\mathbf{b}_j = \begin{bmatrix}\mathbf{w}_1 \cdots \mathbf{w}_m\end{bmatrix}\mathbf{b}_j$ for every $j \in \{1, \ldots, m\}$. Show that $\mathbf{v}_j$ must equal $\mathbf{w}_j$ for every $j \in \{1, \ldots, m\}$.

*Solution* We will first show that $\begin{bmatrix}\mathbf{v}_1 \cdots \mathbf{v}_m\end{bmatrix}$ and $\begin{bmatrix}\mathbf{w}_1 \cdots \mathbf{w}_m\end{bmatrix}$ must have the exact same behavior on every vector in $\mathcal{F}^m$ (where $\mathcal{F}$ is the scalar field) by representing an arbitrary vector $\mathbf{x}$ with respect to $U$. Letting $\mathbf{x} = a_1\mathbf{b}_1 + \ldots + a_m\mathbf{b}_m$,

$$\begin{aligned}
\begin{bmatrix}\mathbf{v}_1 \cdots \mathbf{v}_m\end{bmatrix}\mathbf{x} &= \begin{bmatrix}\mathbf{v}_1 \cdots \mathbf{v}_m\end{bmatrix}(a_1\mathbf{b}_1 + \ldots + a_m\mathbf{b}_m) \\
&= a_1\begin{bmatrix}\mathbf{v}_1 \cdots \mathbf{v}_m\end{bmatrix}\mathbf{b}_1 + \ldots + a_m\begin{bmatrix}\mathbf{v}_1 \cdots \mathbf{v}_m\end{bmatrix}\mathbf{b}_m \\
&= a_1\begin{bmatrix}\mathbf{w}_1 \cdots \mathbf{w}_m\end{bmatrix}\mathbf{b}_1 + \ldots + a_m\begin{bmatrix}\mathbf{w}_1 \cdots \mathbf{w}_m\end{bmatrix}\mathbf{b}_m \\
&= \begin{bmatrix}\mathbf{w}_1 \cdots \mathbf{w}_m\end{bmatrix}(a_1\mathbf{b}_1 + \ldots + a_m\mathbf{b}_m) \\
&= \begin{bmatrix}\mathbf{w}_1 \cdots \mathbf{w}_m\end{bmatrix}\mathbf{x}.
\end{aligned}$$

In particular, the fact that $\begin{bmatrix}\mathbf{v}_1 \cdots \mathbf{v}_m\end{bmatrix}$ and $\begin{bmatrix}\mathbf{w}_1 \cdots \mathbf{w}_m\end{bmatrix}$ map $(1, 0, \ldots, 0)$ to the same vector means that $\mathbf{v}_1$ must equal $\mathbf{w}_1$. Such an argument holds for every *column* in turn. ♦

## 1.6 Eigenvectors

**Definition (Eigenvalue)** Let $\mathbb{T}$ be a linear operator from a vector space to itself. If $\mathbb{T}\mathbf{q} = \lambda\mathbf{q}$ for some scalar $\lambda$ and a non-zero vector $\mathbf{q}$, then $\lambda$ is called an **eigenvalue** of $\mathbb{T}$, and $\mathbf{q}$ is called an **eigenvector** of $\mathbb{T}$ (with eigenvalue $\lambda$). The set of *all* vectors $\mathbf{q}$ that satisfy $\mathbb{T}\mathbf{q} = \lambda\mathbf{q}$ is called the **eigenspace** of $\lambda$ (for $\mathbb{T}$).

**Definition (Principal Eigenvector)** Any unit eigenvector corresponding to the largest eigenvalue of a linear operator is called a **principal eigenvector** of that operator.

**Exercise 1.19** Let $\lambda$ be an eigenvalue for $\mathbb{T}$. Show that the *eigenspace* of $\lambda$ is a *subspace*.

*Solution* Suppose that $\mathbf{q}_1$ and $\mathbf{q}_2$ are both in the eigenspace. For any scalars $a_1, a_2$,

$$\begin{aligned}
\mathbb{T}(a_1\mathbf{q}_1 + a_2\mathbf{q}_2) &= a_1\mathbb{T}\mathbf{q}_1 + a_2\mathbb{T}\mathbf{q}_2 \\
&= a_1\lambda\mathbf{q}_1 + a_2\lambda\mathbf{q}_2 \\
&= \lambda(a_1\mathbf{q}_1 + a_2\mathbf{q}_2)
\end{aligned}$$

which confirms that $a_1\mathbf{q}_1 + a_2\mathbf{q}_2$ is also an eigenvector for $\mathbb{T}$ with eigenvalue $\lambda$. ♦

**Exercise 1.20** Suppose $\mathbb{T}$ has eigenvalues $\lambda_1, \ldots, \lambda_m$ with corresponding eigenvectors $\mathbf{q}_1, \ldots, \mathbf{q}_m$. Let $a$ be a non-zero scalar. Identify eigenvalues and eigenvectors of $a\mathbb{T}$, i.e. the function that maps any vector $\mathbf{v}$ to $a$ times $\mathbb{T}\mathbf{v}$.

*Solution* Consider the action of $a\mathbb{T}$ on $\mathbf{q}_j$.

$$\begin{aligned}[a\mathbb{T}](\mathbf{q}_j) &= a(\mathbb{T}\mathbf{q}_j) \\ &= a\lambda_j\mathbf{q}_j\end{aligned}$$

So $\mathbf{q}_1, \ldots, \mathbf{q}_m$ remain eigenvectors, and their eigenvalues are $a\lambda_1, \ldots, a\lambda_m$. Furthermore, no additional eigenvectors for $a\mathbb{T}$ are introduced because clearly they would also have been eigenvectors for $\mathbb{T}$. ♦

$\mathbb{T}$ is called *invertible* if it has an *inverse*, a function (denoted $\mathbb{T}^{-1}$) that maps each $\mathbb{T}\mathbf{v}$ in $\mathbb{T}$'s range back to $\mathbf{v}$ in $\mathbb{T}$'s domain. The *composition* of $\mathbb{T}^{-1}$ with $\mathbb{T}$ is the *identity* function (denoted $\mathbb{I}$) which maps every element to itself.

**Exercise 1.21** Explain why any linear operator that has 0 as an eigenvalue does not have an inverse function.

*Solution* The corresponding eigenspace is a subspace (with dimension at least 1) that the linear operator maps to $\mathbf{0}$. Because it maps multiple elements of its domain to the same value, it cannot be invertible. ♦

**Exercise 1.22** Let $\mathbb{T}^{-1}$ be the inverse of a linear operator $\mathbb{T}$, that is, $\mathbb{T}^{-1}\mathbb{T}\mathbf{v} = \mathbf{v}$ for every $\mathbf{v}$ in the domain of $\mathbb{T}$. Show that $\mathbb{T}$ is also the inverse of $\mathbb{T}^{-1}$, that is, $\mathbb{T}\mathbb{T}^{-1}\mathbf{w} = \mathbf{w}$ for every $\mathbf{w}$ in the domain of $\mathbb{T}^{-1}$ (which we have defined to be the range of $\mathbb{T}$).

**Exercise 1.23** Suppose a linear operator $\mathbb{T}$ has an inverse $\mathbb{T}^{-1}$. Show that $\mathbb{T}^{-1}$ is also linear:

$$\mathbb{T}^{-1}(a_1\mathbf{w}_1 + a_2\mathbf{w}_2) = a_1\mathbb{T}^{-1}\mathbf{w}_1 + a_2\mathbb{T}^{-1}\mathbf{w}_2$$

for all vectors $\mathbf{w}_1, \mathbf{w}_2$ and scalars $a_1, a_2$.

**Exercise 1.24** Let $\mathbb{T}$ be a linear operator that has non-zero eigenvalues $\lambda_1, \ldots, \lambda_n$ with eigenvectors $\mathbf{q}_1, \ldots, \mathbf{q}_n$. Suppose $\mathbb{T}$ is invertible. Show that $\mathbb{T}^{-1}$ also has $\mathbf{q}_1, \ldots, \mathbf{q}_n$ as eigenvectors, and find the corresponding eigenvalues.

*Solution* Consider the behavior of the inverse on $\mathbf{q}_j$. We know that the inverse is supposed to undo the behavior of $\mathbb{T}$, so $\mathbb{T}^{-1}\mathbb{T}\mathbf{q}_j$ should equal $\mathbf{q}_j$.

$$\begin{aligned}\mathbb{T}^{-1}\mathbb{T}\mathbf{q}_j &= \mathbb{T}^{-1}(\lambda_j\mathbf{q}_j) \\ &= \lambda_j\mathbb{T}^{-1}\mathbf{q}_j\end{aligned}$$

For $\lambda_j \mathbb{T}^{-1}\mathbf{q}_j$ to equal $\mathbf{q}_j$, we can see that $\mathbf{q}_j$ must be an eigenvector of $\mathbb{T}^{-1}$ with eigenvalue $1/\lambda_j$. Thus $\mathbb{T}^{-1}$ has eigenvalues $1/\lambda_1, \ldots, 1/\lambda_n$ with eigenvectors $\mathbf{q}_1, \ldots, \mathbf{q}_n$. ♦

## 1.7 Inner Product

The *dot product* between two vectors in Euclidean space is defined to be the sum of the products of their corresponding coordinates. Many other vector spaces have this same type of operator, providing them with a *geometry* much like that of Euclidean space. The following generalization of *dot product* has a more abstract definition, but it turns out to be essentially equivalent to a sum of products of coordinates as you will see begin to see in Exercise 1.36.[4]

**Definition (Inner Product)** Let $\mathcal{V}$ be a vector space with a scalar field $\mathcal{F}$ that is either the real numbers or the complex numbers. An **inner product** is a function $\langle \cdot, \cdot \rangle$ that maps every pair of vectors to $\mathcal{F}$ and has the following four properties.

- *homogeneity in first argument*: $\langle a\mathbf{v}, \mathbf{w}\rangle = a\langle \mathbf{v}, \mathbf{w}\rangle$ for every $\mathbf{v}, \mathbf{w} \in \mathcal{V}$ and every scalar $a$
- *additivity in first argument*: $\langle \mathbf{v} + \mathbf{w}, \mathbf{x}\rangle = \langle \mathbf{v}, \mathbf{x}\rangle + \langle \mathbf{w}, \mathbf{x}\rangle$ for every $\mathbf{v}, \mathbf{w}, \mathbf{x} \in \mathcal{V}$ and scalar $a$
- *positive definiteness*: $\langle \mathbf{v}, \mathbf{v}\rangle \geq 0$ for every $\mathbf{v} \in \mathcal{V}$ with equality if and only if $\mathbf{v} = \mathbf{0}$
- *conjugate symmetry*: $\langle \mathbf{v}, \mathbf{w}\rangle = \overline{\langle \mathbf{w}, \mathbf{v}\rangle}$ for every $\mathbf{v}, \mathbf{w} \in \mathcal{V}$

**Exercise 1.25** Show that inner products are also additive in their second argument:

$$\langle \mathbf{v}, \mathbf{w} + \mathbf{x}\rangle = \langle \mathbf{v}, \mathbf{w}\rangle + \langle \mathbf{v}, \mathbf{x}\rangle.$$

**Exercise 1.26** Show that inner products are also homogeneous in their second argument when the scalar is real: for every $a \in \mathbb{R}$,

$$\langle \mathbf{v}, a\mathbf{w}\rangle = a\langle \mathbf{v}, \mathbf{w}\rangle.$$

Do not be alarmed by the appearance of a *complex conjugate* in the above definition. For *real* vector spaces, conjugate symmetry simplifies to ordinary symmetry: $\langle \mathbf{v}, \mathbf{w}\rangle = \langle \mathbf{w}, \mathbf{v}\rangle$. From here on, we will strategically avoid having to make further reference to complex conjugates.[5]

**Definition (Orthogonal)** Vectors $\mathbf{v}_1$ and $\mathbf{v}_2$ are called **orthogonal** (denoted $\mathbf{v}_1 \perp \mathbf{v}_2$) if their inner product is zero.

---

[4]For a good discussion of this equivalence in its full generality, see Thm 22.56 of Schechter (1997).

[5]For full disclosure, complex conjugation does appear in the solutions to a few green exercises.

Realize that it is possible for more than one *inner product* operator to be defined on the same vector space, so two vectors might be *orthogonal* with respect to one inner product but not with respect to another. Usually the inner product in question is clear from the context and does not need to be specified explicitly.

The term *orthogonal* is often applied more generally than the above definition indicates. A vector is called *orthogonal* to a subspace if it is orthogonal to every vector in that subspace. Two subspaces are called *orthogonal* to each other if every vector in the first subspace is orthogonal to every vector in the second subspace.

**Exercise 1.27** Show that if $\mathbf{y}$ is orthogonal to every one of $\mathbf{v}_1, \ldots, \mathbf{v}_m$, then it is also orthogonal to every vector in their span.

*Solution* Let $b_1\mathbf{v}_1 + \ldots b_m\mathbf{v}_m$ represent an arbitrary vector in the span. By linearity of inner products, its inner product with $\mathbf{y}$ is

$$\begin{aligned}\langle b_1\mathbf{v}_1 + \ldots b_m\mathbf{v}_m, \mathbf{y}\rangle &= b_1 \underbrace{\langle \mathbf{v}_1, \mathbf{y}\rangle}_{0} + \ldots + b_m \underbrace{\langle \mathbf{v}_m, \mathbf{y}\rangle}_{0} \\ &= 0\end{aligned}$$

because $\mathbf{y}$ is orthogonal to each of the basis vectors. ♦

**Definition (Norm)** Given a vector space along with an inner product, the **norm** of a vector $\mathbf{v}$ is the square root of the inner product of $\mathbf{v}$ with itself:

$$\|\mathbf{v}\| := \sqrt{\langle \mathbf{v}, \mathbf{v}\rangle}.$$

**Exercise 1.28** Show that $\|a\mathbf{v}\| = |a|\|\mathbf{v}\|$ for any scalar $a$.

**Exercise 1.29** Suppose $\mathbf{v}_1, \ldots, \mathbf{v}_m$ are orthogonal to each other and none of them is the zero vector. Show that they must be linearly independent.

**Exercise 1.30** Suppose $\langle \mathbf{x}, \mathbf{v}\rangle = \langle \mathbf{y}, \mathbf{v}\rangle$ for all $\mathbf{v}$. Show that $\mathbf{x}$ and $\mathbf{y}$ must be the same vector.

**Theorem 1.1 (Pythagorean Identity)** *If $\mathbf{v}_1$ and $\mathbf{v}_2$ are orthogonal to each other, then $\|\mathbf{v}_1 + \mathbf{v}_2\|^2 = \|\mathbf{v}_1\|^2 + \|\mathbf{v}_2\|^2$.*

***Proof*** Invoking linearity of inner product to break up the terms,

$$\begin{aligned}\|\mathbf{v}_1 + \mathbf{v}_2\|^2 &= \langle \mathbf{v}_1 + \mathbf{v}_2, \mathbf{v}_1 + \mathbf{v}_2\rangle \\ &= \langle \mathbf{v}_1, \mathbf{v}_1\rangle + \underbrace{\langle \mathbf{v}_1, \mathbf{v}_2\rangle}_{0} + \underbrace{\langle \mathbf{v}_2, \mathbf{v}_1\rangle}_{0} + \langle \mathbf{v}_2, \mathbf{v}_2\rangle \\ &= \|\mathbf{v}_1\|^2 + \|\mathbf{v}_2\|^2.\end{aligned}$$

■

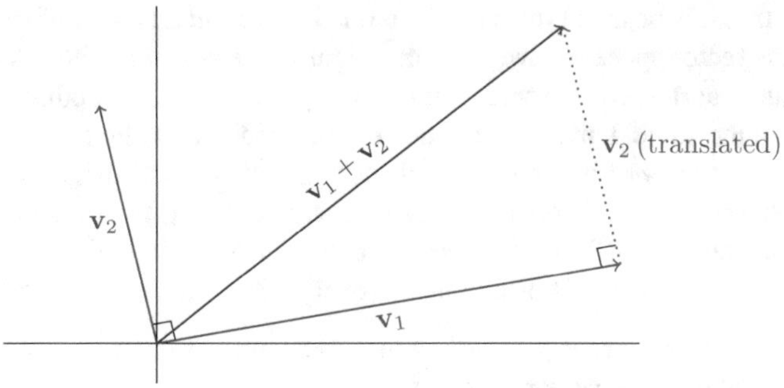


**Fig. 1.2** If $\mathbf{v}_1$ and $\mathbf{v}_2$ are orthogonal, then they can be arranged as sides of a right triangle whose hypotenuse is $\mathbf{v}_1 + \mathbf{v}_2$

To connect this statement of the Pythagorean identity with the usual fact about right triangles, think about vertices at $\mathbf{0}$, $\mathbf{v}_1$, and $\mathbf{v}_1 + \mathbf{v}_2$ forming three sides of a right triangle whose sides are $\mathbf{v}_1$, a translated version of $\mathbf{v}_2$, and $\mathbf{v}_1 + \mathbf{v}_2$, as in Fig. 1.2. Then the Pythagorean identity implies the familiar statement that the squared length of the hypotenuse of a right triangle equals the sum of the squared lengths of the other two sides.

**Exercise 1.31** Justify the Pythagorean identity extended to $m$ orthogonal vectors $\mathbf{v}_1, \ldots, \mathbf{v}_m$:

$$\|\mathbf{v}_1 + \ldots + \mathbf{v}_m\|^2 = \|\mathbf{v}_1\|^2 + \ldots + \|\mathbf{v}_m\|^2.$$

*Solution* $\mathbf{v}_1$ is orthogonal to $\mathbf{v}_2 + \ldots + \mathbf{v}_m$, so by the Pythagorean identity

$$\|\mathbf{v}_1 + \ldots + \mathbf{v}_m\|^2 = \|\mathbf{v}_1\|^2 + \|\mathbf{v}_2 + \ldots + \mathbf{v}_m\|^2.$$

This logic can be applied repeatedly to bring out one vector at a time leading to the desired result. (For a more formal argument, one can invoke *induction*.) ♦

## 1.8 Orthonormal Bases

**Definition (Unit Vector)** Any vector whose norm is 1 is called a **unit vector**.

**Exercise 1.32** Given a non-zero vector $\mathbf{v}$, find the norm of $\frac{1}{\|\mathbf{v}\|}\mathbf{v}$.

*Solution* Using Exercise 1.28 and the fact that norms are non-negative,

$$\begin{aligned}\left\|\frac{1}{\|\mathbf{v}\|}\mathbf{v}\right\| &= \frac{1}{\|\mathbf{v}\|}\|\mathbf{v}\| \\ &= 1.\end{aligned}$$

♦

**Exercise 1.33** Given a unit vector $\mathbf{u}$, find a unique representation of the vector $\mathbf{y}$ as the sum of a vector in the span of $\mathbf{u}$ and a vector orthogonal to the span of $\mathbf{u}$.

*Solution* We will explicitly construct the desired vector in the span of $\mathbf{u}$. The vector we seek must equal $\hat{b}\,\mathbf{u}$ for some scalar $\hat{b}$. Based on the trivial identity $\mathbf{y} = \hat{b}\,\mathbf{u} + (\mathbf{y} - \hat{b}\,\mathbf{u})$, we see that we need the second vector $\mathbf{y} - \hat{b}\,\mathbf{u}$ to be orthogonal to $\mathbf{u}$. Its inner product with $\mathbf{u}$ is

$$\langle \mathbf{y} - \hat{b}\,\mathbf{u}, \mathbf{u} \rangle = \langle \mathbf{y}, \mathbf{u} \rangle - \hat{b} \underbrace{\langle \mathbf{u}, \mathbf{u} \rangle}_{\|\mathbf{u}\|^2=1}$$

which is zero precisely when $\hat{b} = \langle \mathbf{y}, \mathbf{u} \rangle$. Therefore, $\mathbf{y}$ can be represented as the sum of $\langle \mathbf{y}, \mathbf{u} \rangle \mathbf{u}$ which is in the span of $\mathbf{u}$ and $(\mathbf{y} - \langle \mathbf{y}, \mathbf{u} \rangle \mathbf{u})$ which is orthogonal to the span of $\mathbf{u}$. ♦

**Exercise 1.34** Given a non-zero vector $\mathbf{v}$, find a unique representation of the vector $\mathbf{y}$ as the sum of a vector in the span of $\mathbf{v}$ and a vector orthogonal to the span of $\mathbf{v}$.

*Solution* A vector is in the span of $\mathbf{v}$ if and only if it is in the span of the unit vector $\frac{\mathbf{v}}{\|\mathbf{v}\|}$. Likewise, a vector is orthogonal to the span of $\mathbf{v}$ if and only if it is orthogonal to the unit vector $\frac{\mathbf{v}}{\|\mathbf{v}\|}$. Based on our solution to Exercise 1.33 the part in the span of $\mathbf{v}$ must be

$$\left\langle \mathbf{y}, \frac{\mathbf{v}}{\|\mathbf{v}\|} \right\rangle \frac{\mathbf{v}}{\|\mathbf{v}\|} = \frac{\langle \mathbf{y}, \mathbf{v} \rangle}{\|\mathbf{v}\|^2} \mathbf{v}.$$

Thus the desired representation of $\mathbf{y}$ is

$$\mathbf{y} = \underbrace{\frac{\langle \mathbf{y}, \mathbf{v} \rangle}{\|\mathbf{v}\|^2} \mathbf{v}}_{\in \text{span}\{\mathbf{v}\}} + \underbrace{\left( \mathbf{y} - \frac{\langle \mathbf{y}, \mathbf{v} \rangle}{\|\mathbf{v}\|^2} \mathbf{v} \right)}_{\perp \text{span}\{\mathbf{v}\}}.$$

♦

**Definition (Orthonormal)** A set of unit vectors that are all orthogonal to each other are called **orthonormal**. If orthonormal vectors $\mathbf{u}_1, \ldots, \mathbf{u}_m$ span $\mathcal{V}$, the set $\{\mathbf{u}_1, \ldots, \mathbf{u}_m\}$ is called, naturally, an **orthonormal basis** for $\mathcal{V}$.

**Exercise 1.35** Let $\{\mathbf{u}_1, \ldots, \mathbf{u}_m\}$ be an orthonormal basis for $\mathcal{V}$. Find a unique representation of $\mathbf{y} \in \mathcal{V}$ as a linear combination of the basis vectors.

*Solution* The correct coefficients can be readily determined thanks to the orthogonality of the terms:

$$\mathbf{y} = \underbrace{\hat{b}_1 \mathbf{u}_1}_{\in \text{span}\{\mathbf{u}_1\}} + \underbrace{\hat{b}_2 \mathbf{u}_2 + \ldots + \hat{b}_m \mathbf{u}_m}_{\perp \text{span}\{\mathbf{u}_1\}}.$$

By comparison to Exercise 1.33, the first term has to be $\langle \mathbf{y}, \mathbf{u}_1 \rangle \mathbf{u}_1$, so its coefficient has to be $\hat{b}_1 = \langle \mathbf{y}, \mathbf{u}_1 \rangle$. By reasoning similarly for each of the basis vectors, we conclude that $\mathbf{y}$ must have the unique representation

$$\mathbf{y} = \langle \mathbf{y}, \mathbf{u}_1 \rangle \mathbf{u}_1 + \ldots + \langle \mathbf{y}, \mathbf{u}_m \rangle \mathbf{u}_m.$$

♦

Based on Exercise 1.35, any orthonormal basis $\mathbf{u}_1, \ldots, \mathbf{u}_m$ for $\mathcal{V}$ can naturally be understood to provide its own coordinate system for that vector space. The coordinates of $\mathbf{y}$ with respect to $\mathbf{u}_1, \ldots, \mathbf{u}_m$ are $(\langle \mathbf{y}, \mathbf{u}_1 \rangle, \ldots, \langle \mathbf{y}, \mathbf{u}_m \rangle)$. More generally, given any unit vector $\mathbf{u}$, we may also refer to $\langle \mathbf{y}, \mathbf{u} \rangle$ as the *coefficient* or *coordinate* of $\mathbf{y}$ with respect to $\mathbf{u}$.

Let us now verify that the *inner product* and *norm* we have introduced do indeed comport with the definitions that you are accustomed to in the context of Euclidean spaces. The squared norm of $\mathbf{y}$ in a Euclidean space is supposed to be the sum of its squared coordinates with respect to the standard basis; in fact, as Theorem 1.2 verifies, the squared norm, as we have defined it, is also equal to the sum of the squared coordinates with respect to *any* orthonormal basis for a real vector space. More generally, according to Exercise 1.36, the inner product of two vectors, as we have defined it, is equal to the sum of the products of their coordinates with respect to *any* orthonormal basis for a real vector space.

**Exercise 1.36** Let $\{\mathbf{u}_1, \ldots, \mathbf{u}_m\}$ be an orthonormal basis for a real vector space $\mathcal{V}$. Show that the inner product between $\mathbf{x}$ and $\mathbf{y}$ equals the sum of the product of their squared coordinates with respect to $\mathbf{u}_1, \ldots, \mathbf{u}_m$:

$$\langle \mathbf{x}, \mathbf{y} \rangle = \sum_i (\langle \mathbf{x}, \mathbf{u}_i \rangle \langle \mathbf{y}, \mathbf{u}_i \rangle).$$

*Solution* We will use the orthonormal basis representation (Exercise 1.35) to expand $\mathbf{y}$ then use linearity of inner products.

$$\begin{aligned} \langle \mathbf{x}, \mathbf{y} \rangle &= \langle \mathbf{x}, \langle \mathbf{y}, \mathbf{u}_1 \rangle \mathbf{u}_1 + \ldots + \langle \mathbf{y}, \mathbf{u}_m \rangle \mathbf{u}_m \rangle \\ &= \langle \mathbf{x}, \langle \mathbf{y}, \mathbf{u}_1 \rangle \mathbf{u}_1 \rangle + \ldots + \langle \mathbf{x}, \langle \mathbf{y}, \mathbf{u}_m \rangle \mathbf{u}_m \rangle \\ &= \langle \mathbf{y}, \mathbf{u}_1 \rangle \langle \mathbf{x}, \mathbf{u}_1 \rangle + \ldots + \langle \mathbf{y}, \mathbf{u}_m \rangle \langle \mathbf{x}, \mathbf{u}_m \rangle. \end{aligned}$$

♦

**Theorem 1.2 (Parseval's Identity)** *Let* $\{\mathbf{u}_1, \ldots, \mathbf{u}_m\}$ *be any orthonormal basis for a real vector space* $\mathcal{V}$*, and let* $\mathbf{y} \in \mathcal{V}$*. The squared norm of* $\mathbf{y}$ *equals the sum of its squared coordinates with respect to* $\mathbf{u}_1, \ldots, \mathbf{u}_m$

*Proof* According to Exercise 1.36, the inner product of any two vectors is equal to the sum of the products of their coordinates with respect to $\mathbf{u}_1, \ldots, \mathbf{u}_m$. The squared norm of $\mathbf{y}$ is its inner product with itself, so

$$\begin{aligned}\|\mathbf{y}\|^2 &= \langle \mathbf{y}, \mathbf{y}\rangle \\ &= \sum_i (\langle \mathbf{y}, \mathbf{u}_i\rangle \langle \mathbf{y}, \mathbf{u}_i\rangle).\end{aligned}$$

■

**Exercise 1.37** Let $\{\mathbf{u}_1, \ldots, \mathbf{u}_m\}$ be an orthonormal basis for a real vector space $\mathcal{V}$, and let $\mathbf{y} \in \mathcal{V}$. Consider the *approximation* $\hat{\mathbf{y}} := \langle \mathbf{y}, \mathbf{u}_1\rangle \mathbf{u}_1 + \ldots + \langle \mathbf{y}, \mathbf{u}_k\rangle \mathbf{u}_k$ with $k \leq m$. Use Parseval's identity to derive a simple formula for the squared norm of $\mathbf{y} - \hat{\mathbf{y}}$, which we might call the *squared approximation error*.

*Solution* Representing $\mathbf{y}$ with respect to the orthonormal basis, we find that the difference between the vectors is

$$\begin{aligned}\mathbf{y} - \hat{\mathbf{y}} &= (\langle \mathbf{y}, \mathbf{u}_1\rangle \mathbf{u}_1 + \ldots + \langle \mathbf{y}, \mathbf{u}_m\rangle \mathbf{u}_m) - (\langle \mathbf{y}, \mathbf{u}_1\rangle \mathbf{u}_1 + \ldots + \langle \mathbf{y}, \mathbf{u}_k\rangle \mathbf{u}_k) \\ &= \langle \mathbf{y}, \mathbf{u}_{k+1}\rangle \mathbf{u}_{k+1} + \ldots + \langle \mathbf{y}, \mathbf{u}_m\rangle \mathbf{u}_m.\end{aligned}$$

Its squared norm is the sum of its squared coordinates, so

$$\|\mathbf{y} - \hat{\mathbf{y}}\|^2 = \langle \mathbf{y}, \mathbf{u}_{k+1}\rangle^2 + \ldots + \langle \mathbf{y}, \mathbf{u}_m\rangle^2.$$

♦

**Exercise 1.38** Let $\{\mathbf{u}_1, \ldots, \mathbf{u}_m\}$ be an orthonormal basis for a real vector space $\mathcal{V}$, and let $\mathbf{y} \in \mathcal{V}$. Explain which term in the representation $\langle \mathbf{y}, \mathbf{u}_1\rangle \mathbf{u}_1 + \ldots + \langle \mathbf{y}, \mathbf{u}_m\rangle \mathbf{u}_m$ best approximates $\mathbf{y}$ in the sense that it results in the smallest approximation error $\|\mathbf{y} - \langle \mathbf{y}, \mathbf{u}_j\rangle \mathbf{u}_j\|$.

*Solution* Based on Exercise 1.37, the squared approximation error $\|\mathbf{y} - \langle \mathbf{y}, \mathbf{u}_j\rangle \mathbf{u}_j\|^2$ is equal to the sum of the squares of the other coefficients $\sum_{i \neq j} \langle \mathbf{y}, \mathbf{u}_i\rangle^2$. Therefore, the approximation error is minimized if we use the term with the largest squared coefficient. ♦

Given any basis $\{\mathbf{v}_1, \ldots, \mathbf{v}_m\}$ for $\mathcal{V}$, the *Gram–Schmidt algorithm* is a process for constructing an orthonormal basis $\{\mathbf{u}_1, \ldots, \mathbf{u}_m\}$ from the original basis vectors. First, *normalize* the first basis vector

$$\mathbf{u}_1 := \frac{\mathbf{v}_1}{\|\mathbf{v}_1\|}.$$

Then $\mathbf{v}_2 - \langle \mathbf{v}_2, \mathbf{u}_1\rangle \mathbf{u}_1$ is orthogonal to $\mathbf{u}_1$; dividing by its length creates a vector

$$\mathbf{u}_2 := \frac{\mathbf{v}_2 - \langle \mathbf{v}_2, \mathbf{u}_1\rangle \mathbf{u}_1}{\|\mathbf{v}_2 - \langle \mathbf{v}_2, \mathbf{u}_1\rangle \mathbf{u}_1\|}$$

which remains orthogonal to $\mathbf{u}_1$ but has length 1. We know from Exercise 1.4 that $\mathbf{u}_1$ and $\mathbf{u}_2$ have the same span as $\mathbf{v}_1$ and $\mathbf{v}_2$. (Notice that $\mathbf{v}_2 - \langle \mathbf{v}_2, \mathbf{u}_1\rangle \mathbf{u}_1$ is guaranteed to have positive length because it can be represented as a non-zero linear combination of $\mathbf{v}_1$ and $\mathbf{v}_2$ and therefore cannot be the zero vector.) Repeat this process to define

$$\mathbf{u}_3 := \frac{\mathbf{v}_3 - (\langle \mathbf{v}_3, \mathbf{u}_1 \rangle \mathbf{u}_1 + \langle \mathbf{v}_3, \mathbf{u}_2 \rangle \mathbf{u}_2)}{\|\mathbf{v}_3 - (\langle \mathbf{v}_3, \mathbf{u}_1 \rangle \mathbf{u}_1 + \langle \mathbf{v}_3, \mathbf{u}_2 \rangle \mathbf{u}_2)\|}$$

$$\vdots$$

$$\mathbf{u}_m := \frac{\mathbf{v}_m - (\langle \mathbf{v}_m, \mathbf{u}_1 \rangle \mathbf{u}_1 + \ldots + \langle \mathbf{v}_m, \mathbf{u}_{m-1} \rangle \mathbf{u}_{m-1})}{\|\mathbf{v}_m - (\langle \mathbf{v}_m, \mathbf{u}_1 \rangle \mathbf{u}_1 + \ldots + \langle \mathbf{v}_m, \mathbf{u}_{m-1} \rangle \mathbf{u}_{m-1})\|}$$

in turn. These $m$ orthonormal vectors must be a basis for $\mathcal{V}$ because it has dimension $m$.

## 1.9 Orthogonal Projection

**Definition (Orthogonal Complement)** Let $\mathcal{S}$ be a subspace of a vector space $\mathcal{V}$. Given a subspace $\mathcal{S}$, the set of all vectors that are orthogonal to $\mathcal{S}$ is called its **orthogonal complement** and is denoted $\mathcal{S}^{\perp}$. If $\mathcal{W}$ is another subspace of $\mathcal{V}$ such that $\mathcal{S} \subseteq \mathcal{W}$, the set of all vectors in $\mathcal{W}$ that are orthogonal to $\mathcal{S}$ is called the **orthogonal complement** of $\mathcal{S}$ within $\mathcal{W}$.

**Exercise 1.39** Given a subspace $\mathcal{S}$, show that $\mathcal{S}^{\perp}$ is also a subspace.

*Solution* Let $\mathbf{v}_1, \mathbf{v}_2 \in \mathcal{S}^{\perp}$, and let $b_1$ and $b_2$ be scalars. We need to show that the linear combination $b_1\mathbf{v}_1 + b_2\mathbf{v}_2$ is also in $\mathcal{S}^{\perp}$. Letting $\mathbf{w}$ be an arbitrary vector in $\mathcal{S}$,

$$\begin{aligned} \langle b_1\mathbf{v}_1 + b_2\mathbf{v}_2, \mathbf{w} \rangle &= b_1 \underbrace{\langle \mathbf{v}_1, \mathbf{w} \rangle}_{0} + b_2 \underbrace{\langle \mathbf{v}_2, \mathbf{w} \rangle}_{0} \\ &= 0. \end{aligned}$$

♦

**Definition (Orthogonal Projection)** Let $\mathbf{y} \in \mathcal{V}$ be a vector and let $\mathcal{S}$ be a subspace of $\mathcal{V}$. If there exists $\hat{\mathbf{y}} \in \mathcal{S}$ such that $\mathbf{y} - \hat{\mathbf{y}} \perp \mathcal{S}$, then $\hat{\mathbf{y}}$ is the **orthogonal projection** of $\mathbf{y}$ onto $\mathcal{S}$.

Is an orthogonal projection guaranteed to exist? And when it does exist, is it unique? Those are important questions, and the following exercises will help us answer them.

**Exercise 1.40** Let $\hat{\mathbf{y}}$ be the orthogonal projection of $\mathbf{y}$ onto $\mathcal{S}$. Use the Pythagorean identity to show that the vector in $\mathcal{S}$ that is closest to $\mathbf{y}$ is $\hat{\mathbf{y}}$.

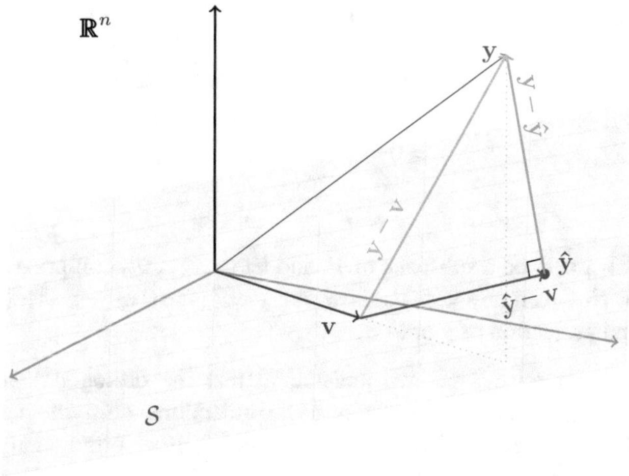


**Fig. 1.3** The right triangle in the picture clarifies that the orthogonal projection is the vector in $\mathcal{S}$ that is closest to $\mathbf{y}$. According to the Pythagorean identity, the squared distance from $\mathbf{y}$ to any $\mathbf{v} \in \mathcal{S}$ is the squared distance from $\mathbf{y}$ to $\hat{\mathbf{y}}$ plus the squared distance from $\hat{\mathbf{y}}$ to $\mathbf{v}$

*Solution* Let $\mathbf{v}$ be an arbitrary vector in $\mathcal{S}$. Realizing that $\hat{\mathbf{y}} - \mathbf{v}$ is in $\mathcal{S}$ and that $\mathbf{y} - \hat{\mathbf{y}}$ is orthogonal to $\mathcal{S}$, we observe a right triangle (Fig. 1.3) with sides $\mathbf{y} - \mathbf{v}$, $\hat{\mathbf{y}} - \mathbf{v}$, and $\mathbf{y} - \hat{\mathbf{y}}$. By the Pythagorean identity,

$$\|\mathbf{y} - \mathbf{v}\|^2 = \|\mathbf{y} - \hat{\mathbf{y}}\|^2 + \|\hat{\mathbf{y}} - \mathbf{v}\|^2.$$

The first term on the right does not depend on the choice of $\mathbf{v}$, so the quantity is uniquely minimized by choosing $\mathbf{v}$ equal to $\hat{\mathbf{y}}$ to make the second term zero. ♦

If an orthogonal projection exists, it is the uniquely closest vector in the subspace according to Exercise 1.40. There cannot be more than one *uniquely* closest vector, so orthogonal projections must be *unique*.

**Exercise 1.41** Let $\mathcal{S}_1$ and $\mathcal{S}_2$ be subspaces that are orthogonal to each other, and let $\mathcal{S}$ be the span of their union. If $\hat{\mathbf{y}}_1$ and $\hat{\mathbf{y}}_2$ are the orthogonal projections of $\mathbf{y}$ onto $\mathcal{S}_1$ and $\mathcal{S}_2$, show that the orthogonal projection of $\mathbf{y}$ onto $\mathcal{S}$ is $\hat{\mathbf{y}}_1 + \hat{\mathbf{y}}_2$.

*Solution* For an arbitrary $\mathbf{v} \in \mathcal{S}$, we need to establish that

$$\mathbf{y} - (\hat{\mathbf{y}}_1 + \hat{\mathbf{y}}_2) \perp \mathbf{v}.$$

Every vector in the span of $\mathcal{S}_1 \cup \mathcal{S}_2$ can be represented as the sum of a vector in $\mathcal{S}_1$ and a vector in $\mathcal{S}_2$. Making use of this fact, we let $\mathbf{v} = \mathbf{v}_1 + \mathbf{v}_2$ with $\mathbf{v}_1 \in \mathcal{S}_1$ and $\mathbf{v}_2 \in \mathcal{S}_2$.

$$\langle \mathbf{v}, \mathbf{y} - (\hat{\mathbf{y}}_1 + \hat{\mathbf{y}}_2) \rangle = \langle \mathbf{v}_1 + \mathbf{v}_2, \mathbf{y} - \hat{\mathbf{y}}_1 - \hat{\mathbf{y}}_2 \rangle$$

$$= \langle \mathbf{v}_1, \mathbf{y} - \hat{\mathbf{y}}_1 - \hat{\mathbf{y}}_2 \rangle + \langle \mathbf{v}_2, \mathbf{y} - \hat{\mathbf{y}}_1 - \hat{\mathbf{y}}_2 \rangle$$
$$= \underbrace{\langle \mathbf{v}_1, \mathbf{y} - \hat{\mathbf{y}}_1 \rangle}_{0} + \underbrace{\langle \mathbf{v}_2, \mathbf{y} - \hat{\mathbf{y}}_2 \rangle}_{0}$$
$$= 0$$

♦

**Exercise 1.42** Let $\mathcal{S}$ be a subspace of $\mathcal{V}$, and let $\mathbf{u}_1, \ldots, \mathbf{u}_m$ comprise an orthonormal basis for $\mathcal{S}$. Given any $\mathbf{y} \in \mathcal{V}$, show that $\hat{\mathbf{y}} := \langle \mathbf{y}, \mathbf{u}_1 \rangle \mathbf{u}_1 + \ldots + \langle \mathbf{y}, \mathbf{u}_m \rangle \mathbf{u}_m$ is the orthogonal projection of $\mathbf{y}$ onto $\mathcal{S}$.

*Solution* From Exercise 1.41, we understand that the orthogonal projection of onto $\mathcal{S}$ equals the sum of its orthogonal projections onto the spans of the orthonormal basis vectors. The representations of these orthogonal projections as $\langle \mathbf{y}, \mathbf{u}_1 \rangle \mathbf{u}_1, \ldots, \langle \mathbf{y}, \mathbf{u}_m \rangle \mathbf{u}_m$ come from Exercise 1.33. ♦

The formula in Exercise 1.42 makes it clear that the orthogonal projection of any vector onto a finite-dimensional subspace *exists*.[6]

Here's another way to think about finding orthogonal projections. If you obtain a representation $\mathbf{y} = \mathbf{v}_1 + \mathbf{v}_2$ such that $\mathbf{v}_1 \in \mathcal{S}$ and $\mathbf{v}_2 \perp \mathcal{S}$, then $\mathbf{v}_1$ is the orthogonal projection of $\mathbf{y}$ onto $\mathcal{S}$. (Just rearrange the equation to see this method's connection to the definition: $\mathbf{v}_2 = \mathbf{y} - \mathbf{v}_1$.)

**Exercise 1.43** Suppose $\hat{\mathbf{y}}_1$ and $\hat{\mathbf{y}}_2$ are the orthogonal projections of $\mathbf{y}_1$ and $\mathbf{y}_2$ onto $\mathcal{S}$. With scalars $a_1$ and $a_2$, find the orthogonal projection of $a_1\mathbf{y}_1 + a_2\mathbf{y}_2$ onto $\mathcal{S}$.

*Solution* We can write out each vector in terms of its orthogonal projections onto $\mathcal{S}$ and $\mathcal{S}^\perp$, then regroup the terms.

$$a_1\mathbf{y}_1 + a_2\mathbf{y}_2 = a_1[\hat{\mathbf{y}}_1 + (\mathbf{y}_1 - \hat{\mathbf{y}}_1)] + a_2[\hat{\mathbf{y}}_2 + (\mathbf{y}_2 - \hat{\mathbf{y}}_2)]$$
$$= \underbrace{(a_1\hat{\mathbf{y}}_1 + a_2\hat{\mathbf{y}}_2)}_{\in\mathcal{S}} + \underbrace{[a_1(\mathbf{y}_1 - \hat{\mathbf{y}}_1) + a_2(\mathbf{y}_2 - \hat{\mathbf{y}}_2)]}_{\perp\mathcal{S}}$$

This shows that $a_1\hat{\mathbf{y}}_1 + a_2\hat{\mathbf{y}}_2$ is the orthogonal projection of $a_1\mathbf{y}_1 + a_2\mathbf{y}_2$ onto $\mathcal{S}$. In other words, the orthogonal projection of a linear combination is the linear combination of the orthogonal projections. ♦

**Exercise 1.44** Let $\hat{\mathbf{y}}$ be the orthogonal projection of $\mathbf{y}$ onto $\mathcal{S}$. How do we know that $\mathbf{y} - \hat{\mathbf{y}}$ is the orthogonal projection of $\mathbf{y}$ onto $\mathcal{S}^\perp$?

*Solution* We know that $\mathbf{y} = \hat{\mathbf{y}} + (\mathbf{y} - \hat{\mathbf{y}})$ with $\hat{\mathbf{y}} \in \mathcal{S}$ and $\mathbf{y} - \hat{\mathbf{y}} \in \mathcal{S}^\perp$ by definition of orthogonal projection. Of course, by definition of orthogonal complement, $\hat{\mathbf{y}} \perp \mathcal{S}^\perp$,

[6] Infinite-dimensional subspaces can have orthogonal projections too under the right conditions, but the finite-dimensional case is perfectly adequate for this text.

so that same representation shows that $\mathbf{y} - \hat{\mathbf{y}}$ is the orthogonal projection of $\mathbf{y}$ onto $\mathcal{S}^{\perp}$. ♦

**Exercise 1.45** Let $\mathcal{S}$ be an $m$-dimensional subspace of a $d$-dimensional vector space $\mathcal{V}$. Verify that the dimension of $\mathcal{S}^{\perp}$ is $d - m$.

We will often have to deal with orthogonal projections of a vector onto *nested* subspaces $\mathcal{S}_0 \subseteq \mathcal{S}_1$, so let us go ahead and give this situation some thought. Let $\hat{\mathbf{y}}_1$ represent the orthogonal projection of $\mathbf{y}$ onto $\mathcal{S}_1$. Let $\hat{\mathbf{y}}_0$ represent the orthogonal projection of $\hat{\mathbf{y}}_1$ onto $\mathcal{S}_0$. By adding and subtracting these vectors we can represent $\mathbf{y}$ conveniently as

$$\mathbf{y} = \hat{\mathbf{y}}_0 + (\hat{\mathbf{y}}_1 - \hat{\mathbf{y}}_0) + (\mathbf{y} - \hat{\mathbf{y}}_1). \tag{1.4}$$

Let us make a couple of important observations from this equation. The first term $\hat{\mathbf{y}}_0$ is in $\mathcal{S}_0$, while the remaining terms are orthogonal to $\mathcal{S}_0$, meaning that it must be the orthogonal projection of $\mathbf{y}$ onto $\mathcal{S}_0$. Recall that we defined $\hat{\mathbf{y}}_0$ to be the orthogonal projection of $\hat{\mathbf{y}}_1$ onto $\mathcal{S}_0$. In other words, whether you project $\mathbf{y}$ onto $\mathcal{S}_1$ then project that result onto $\mathcal{S}_0$ or you project $\mathbf{y}$ directly onto $\mathcal{S}_0$, you end up at the same point.

Next, let us inspect the middle vector $\hat{\mathbf{y}}_1 - \hat{\mathbf{y}}_0$. It is in $\mathcal{S}_1$, but it is also orthogonal to $\mathcal{S}_0$. In other words, it is in the orthogonal complement of $\mathcal{S}_0$ within $\mathcal{S}_1$. The remaining vectors are orthogonal to this subspace, so their sum $\hat{\mathbf{y}}_0 + (\mathbf{y} - \hat{\mathbf{y}}_1)$ is also orthogonal to it. Therefore, this representation reveals that $\hat{\mathbf{y}}_1 - \hat{\mathbf{y}}_0$ is the orthogonal projection of $\mathbf{y}$ onto the orthogonal complement of $\mathcal{S}_0$ within $\mathcal{S}_1$.

## 1.10 Orthogonal Projection Operators

**Definition (Orthogonal Projection Operator)** A function that maps every vector to its orthogonal projection onto a subspace $\mathcal{S}$ is called the **orthogonal projection operator** onto $\mathcal{S}$.

We have previously observed that the orthogonal projection of any vector onto a finite-dimensional subspace always exists, so we can be assured that orthogonal projection operators exist for such subspaces. Furthermore, because orthogonal projections are unique, any orthogonal projection operator must also be unique.

Based on Exercise 1.43, we can infer that *orthogonal projection operators are linear operators.*

**Exercise 1.46** Let $\mathbb{H}$ be an orthogonal projection operator onto $\mathcal{S}$. Show that every vector in $\mathcal{S}$ is an eigenvector of $\mathbb{H}$.

*Solution* If $\mathbf{v}$ is in $\mathcal{S}$, then clearly $\mathbf{v} = \mathbf{v} + \mathbf{0}$ is the unique representation of $\mathbf{v}$ as the sum of a vector in $\mathcal{S}$ and a vector orthogonal to $\mathcal{S}$. Therefore $\mathbb{H}\mathbf{v} = \mathbf{v}$, which means that $\mathbf{v}$ is an eigenvector with eigenvalue 1. ♦

**Exercise 1.47** Let $\mathbb{H}$ be the orthogonal projection operator onto $\mathcal{S}$. Show that every vector in $\mathcal{S}^\perp$ is an eigenvector of $\mathbb{H}$.

*Solution* If $\mathbf{v} \perp \mathcal{S}$, then clearly $\mathbf{v} = \mathbf{0} + \mathbf{v}$ is the unique representation of $\mathbf{v}$ as the sum of a vector in $\mathcal{S}$ and a vector orthogonal to $\mathcal{S}$. Therefore $\mathbb{H}\mathbf{v} = \mathbf{0}$, which means that $\mathbf{v}$ is an eigenvector with eigenvalue 0. ♦

**Definition (Idempotent)** $\mathbb{T}$ is called **idempotent**[7] if $\mathbb{T} \circ \mathbb{T} = \mathbb{T}$.

**Exercise 1.48** Show that every orthogonal projection operator is idempotent.

*Solution* Let $\mathbb{H}$ be the orthogonal projection operator onto $\mathcal{S}$, and let $\hat{\mathbf{y}}$ be the orthogonal projection of $\mathbf{y}$ onto $\mathcal{S}$. Because $\hat{\mathbf{y}}$ is in $\mathcal{S}$, $\mathbb{H}$ maps it to itself.

$$\begin{aligned}[\mathbb{H} \circ \mathbb{H}]\mathbf{y} &= \mathbb{H}(\mathbb{H}\mathbf{y}) \\ &= \mathbb{H}\hat{\mathbf{y}} \\ &= \hat{\mathbf{y}}\end{aligned}$$

The action of $\mathbb{H} \circ \mathbb{H}$ is exactly the same as that of $\mathbb{H}$ on every vector, so they are the same operator. ♦

**Exercise 1.49** Let $\mathbb{H}_1$ be the orthogonal projection operator onto $\mathcal{S}_1$, and let $\mathbb{H}_0$ be the orthogonal projection operator onto $\mathcal{S}_0 \subseteq \mathcal{S}_1$. Explain why $\mathbb{H}_1 - \mathbb{H}_0$ is the orthogonal projection operator onto the orthogonal complement of $\mathcal{S}_0$ within $\mathcal{S}_1$.

**Exercise 1.50** Let $\mathcal{S}_0 \subseteq \mathcal{S}_1$ be subspaces, and let $\mathbb{H}_0$ and $\mathbb{H}_1$ be orthogonal projection operators onto $\mathcal{S}_0$ and $\mathcal{S}_1$, respectively. Explain why $\mathbb{H}_0 \circ \mathbb{H}_1 = \mathbb{H}_1 \circ \mathbb{H}_0 = \mathbb{H}_0$.

## 1.11 Matrices

**Definition (Matrix)** A rectangular arrangement of elements from a field $\mathcal{F}$ into $n$ rows and $m$ columns is called an $n \times m$ **matrix**. The set of all $n \times m$ matrices of elements of $\mathcal{F}$ is denoted $\mathcal{F}^{n \times m}$.

The *product* of two matrices $\mathbb{M}$ times $\mathbb{V}$ is defined to be the matrix whose $(i, j)$-entry (the entry in row $i$ and column $j$) is the sum of the products of the entries of the $i$th row of $\mathbb{M}$ with the $j$th column of $\mathbb{V}$.

[7] *Idempotent* comes from Latin *idem* (root of *identical*) meaning *the same* and *potentia* meaning *power*. It is an appropriate name because if $\mathbb{T}$ is idempotent, then $\mathbb{T}^k = \mathbb{T}$ for every positive integers $k$. (Here, $\mathbb{T}^k$ means that $\mathbb{T}$ has been composed with itself $k$ times.)

$$[\mathbb{MV}]_{i,j} := \sum_k \mathbb{M}_{i,k}\mathbb{V}_{k,j}$$

Letting $\mathbf{m}_1, \ldots, \mathbf{m}_n$ denote the *rows* of $\mathbb{M}$ and $\mathbf{v}_1, \ldots, \mathbf{v}_m$ denote the *columns* of $\mathbb{V}$, we can also depict their product as

$$\mathbb{MV} = \begin{bmatrix} - & \mathbf{m}_1 & - \\ & \vdots & \\ - & \mathbf{m}_n & - \end{bmatrix} \begin{bmatrix} | & & | \\ \mathbf{v}_1 & \ldots & \mathbf{v}_m \\ | & & | \end{bmatrix} = \begin{bmatrix} \mathbf{m}_1^T\mathbf{v}_1 & \cdots & \mathbf{m}_1^T\mathbf{v}_m \\ \vdots & \ddots & \vdots \\ \mathbf{m}_n^T\mathbf{v}_1 & \cdots & \mathbf{m}_n^T\mathbf{v}_m \end{bmatrix}.$$

Note that each entry is written as the product of a $1 \times m$ matrix with an $m \times 1$ matrix; any such $\mathbf{m}_i^T\mathbf{v}_j$ can be seen to equal the familiar dot product between $\mathbf{m}_i$ and $\mathbf{v}_j$.

In the special case that the matrix on the right side of the multiplication has a single column, we can think of it as a [column] vector of scalars. The product is conveniently represented as the *linear combination of the columns* of the matrix, with the vector's entries providing the coefficients. For instance, multiplying $\mathbb{V} \in \mathcal{F}^{m\times n}$ by $\mathbf{b} = (b_1, \ldots, b_m)$ results in

$$\begin{bmatrix} \mathbb{V}_{1,1} & \cdots & \mathbb{V}_{1,m} \\ \vdots & \ddots & \vdots \\ \mathbb{V}_{n,1} & \cdots & \mathbb{V}_{n,m} \end{bmatrix} \begin{bmatrix} b_1 \\ \vdots \\ b_m \end{bmatrix} = \begin{bmatrix} b_1\mathbb{V}_{1,1} + \ldots + b_m\mathbb{V}_{1,m} \\ \vdots \\ b_1\mathbb{V}_{n,1} + \ldots + b_m\mathbb{V}_{n,m} \end{bmatrix} = b_1 \begin{bmatrix} \mathbb{V}_{1,1} \\ \vdots \\ \mathbb{V}_{n,1} \end{bmatrix} + \ldots + b_m \begin{bmatrix} \mathbb{V}_{1,m} \\ \vdots \\ \mathbb{V}_{n,m} \end{bmatrix}.$$

This is a special case of the notation developed for linear combinations in Eq. 1.1. Our observations in that section establish the following fundamental fact about multiplying $\mathbb{M} \in \mathcal{F}^{n\times m}$ with a vector. The mapping from $\mathbf{b} \in \mathcal{F}^m$ to $\mathbb{M}\mathbf{b}$ is a *linear operator*: for any $\mathbf{b}_1, \mathbf{b}_2 \in \mathcal{F}^m$ and $a_1, a_2 \in \mathcal{F}$,

$$\mathbb{M}(a_1\mathbf{b}_1 + a_2\mathbf{b}_2) = a_1\mathbb{M}\mathbf{b}_1 + a_2\mathbb{M}\mathbf{b}_2.$$

The product of matrices is the same as their *composition* as operators. Function composition is always associative, thus so is matrix multiplication.

The following definitions will refresh your memory regarding some of the basic concepts related to matrices, and the exercises will construct tools for our later use.

**Definition (Column and Row Spaces)** Let $\mathbb{M}$ be a matrix. The **column space** of $\mathbb{M}$, denoted $C(\mathbb{M})$, is the span of its column vectors. The **row space** of $\mathbb{M}$, denoted $R(\mathbb{M})$, is the span of its row vectors.

**Definition (Rank)** The **rank** of a matrix is the dimension of its column space. A matrix whose columns are linearly independent has **full (column) rank**. A matrix whose rows are linearly independent has **full row rank**.

**Definition (Diagonal)** Let $\mathbb{M}$ be an $n \times m$ matrix, and let $d := \min(m, n)$. The **diagonal** of $\mathbb{M}$ comprises its $(1, 1), (2, 2), \ldots, (d, d)$ entries. The other entries of $\mathbb{M}$ are called **off-diagonal**. A matrix itself is called **diagonal** if every off-diagonal entry is zero.

**Definition (Trace)** The **trace** of a square matrix is the sum along its diagonal.

**Exercise 1.51** Let $\mathbb{M} \in \mathcal{F}^{n\times m}$ and $\mathbb{L} \in \mathcal{F}^{m\times n}$. Show that the trace of $\mathbb{ML}$ is equal to the trace of $\mathbb{LM}$.

**Definition (Transpose)** The **transpose** of a matrix $\mathbb{M}$, denoted $\mathbb{M}^T$, is the matrix whose columns are the rows of $\mathbb{M}$; likewise the rows of $\mathbb{M}^T$ are the columns of $\mathbb{M}$.

**Exercise 1.52** Let $\mathbb{M}$ and $\mathbb{L}$ be matrices such that the product $\mathbb{ML}$ is well-defined. Show that $(\mathbb{ML})^T = \mathbb{L}^T\mathbb{M}^T$.

**Definition (Symmetric)** A matrix is called **symmetric** if it is equal to its transpose.

An *identity matrix* (which maps each vector to itself) has a 1 as each of its diagonal entries and a 0 as each of its off-diagonal entries. The notation $\mathbb{I}_n$ is used to explicitly specify the $n \times n$ identity matrix. Notice that an identity matrix must be *square*, meaning it has the same number of rows and columns.

**Exercise 1.53** If the columns of $\mathbb{U}$ are orthonormal, show that $\mathbb{U}^T\mathbb{U}$ equals the identity matrix $\mathbb{I}$.

**Exercise 1.54** Show that $\mathbb{M}^T\mathbb{M}$ is symmetric.

*Solution* The transpose of a product of matrices is equal to the product of their transposes multiplied in the reverse order (Exercise 1.52). Thus

$$\begin{aligned}(\mathbb{M}^T\mathbb{M})^T &= (\mathbb{M})^T(\mathbb{M}^T)^T \\ &= \mathbb{M}^T\mathbb{M}.\end{aligned}$$

♦

**Exercise 1.55** Show that a square matrix is invertible if and only if its columns are linearly independent.

## 1.12 Matrix Decompositions

**Theorem 1.3 (Spectral Theorem, Penney (2015) Sec 6.6)** *Any real symmetric $n \times n$ matrix has $n$ orthonormal eigenvectors with real eigenvalues.*

**Exercise 1.56** Suppose $\mathbb{M} \in \mathbb{R}^{n \times n}$ has orthonormal eigenvectors $\mathbf{q}_1, \ldots, \mathbf{q}_n$ with eigenvalues $\lambda_1, \ldots, \lambda_n$ Show that $\mathbb{M}$ cannot have any other eigenvalues.

**Exercise 1.57** Let $\mathbf{q}_1, \ldots, \mathbf{q}_n$ be an orthonormal basis for $\mathbb{R}^n$. Show that $\mathbb{M}$ has the *spectral decomposition*

$$\mathbb{M} = \lambda_1 \mathbf{q}_1 \mathbf{q}_1^T + \ldots + \lambda_n \mathbf{q}_n \mathbf{q}_n^T$$

if and only if $\mathbf{q}_1, \ldots, \mathbf{q}_n$ are eigenvectors for $\mathbb{M}$ with eigenvalues $\lambda_1, \ldots, \lambda_n$.

*Solution* Let us figure out the behavior of $\lambda_1 \mathbf{q}_1 \mathbf{q}_1^T + \ldots + \lambda_n \mathbf{q}_n \mathbf{q}_n^T$ on the basis vectors.

$$\begin{aligned}(\lambda_1 \mathbf{q}_1 \mathbf{q}_1^T + \ldots + \lambda_n \mathbf{q}_n \mathbf{q}_n^T)\mathbf{q}_1 &= \lambda_1 \mathbf{q}_1 \underbrace{\mathbf{q}_1^T \mathbf{q}_1}_{\|\mathbf{q}_1\|^2=1} + \ldots + \lambda_n \mathbf{q}_n \underbrace{\mathbf{q}_n^T \mathbf{q}_1}_{0} \\ &= \lambda_1 \mathbf{q}_1\end{aligned}$$

meaning $\mathbf{q}_1$ is also an eigenvector of this matrix with eigenvalue $\lambda_1$. Likewise for $\mathbf{q}_2, \ldots, \mathbf{q}_n$. By establishing that $\mathbb{M}$ and $\lambda_1 \mathbf{q}_1 \mathbf{q}_1^T + \ldots + \lambda_n \mathbf{q}_n \mathbf{q}_n^T$ behave the same on a basis, we see that they must be the same matrix by Exercise 1.18. ♦

The behavior of $\mathbb{M}$ is very easily described with respect to the eigenvector basis: it simply multiplies the $j$th coordinate by $\lambda_j$, as a spectral decomposition reveals.

$$\begin{aligned}\mathbb{M}\mathbf{y} &= (\lambda_1 \mathbf{q}_1 \mathbf{q}_1^T + \ldots + \lambda_n \mathbf{q}_n \mathbf{q}_n^T)\mathbf{y} \\ &= \lambda_1 \langle \mathbf{y}, \mathbf{q}_1 \rangle \mathbf{q}_1 + \ldots + \lambda_n \langle \mathbf{y}, \mathbf{q}_n \rangle \mathbf{q}_n \qquad (1.5)\end{aligned}$$

We see from this expression that the coordinates of $\mathbb{M}\mathbf{y}$ with respect the eigenvector basis are simply $\lambda_1, \ldots, \lambda_n$ times the corresponding coordinates of $\mathbf{y}$ with respect to that basis.

**Exercise 1.58** Let $\mathbb{M} \in \mathbb{R}^{n \times n}$ be a symmetric matrix with non-negative eigenvalues $\lambda_1, \ldots, \lambda_n$ and corresponding orthonormal eigenvectors $\mathbf{q}_1, \ldots, \mathbf{q}_n$. Show that the symmetric matrix that has eigenvalues $\sqrt{\lambda_1}, \ldots, \sqrt{\lambda_n}$ with eigenvectors $\mathbf{q}_1, \ldots, \mathbf{q}_n$ is the *square root* of $\mathbb{M}$ (denoted $\mathbb{M}^{1/2}$) in the sense that $\mathbb{M}^{1/2}\mathbb{M}^{1/2} = \mathbb{M}$.

*Solution* Using a spectral decomposition, we multiply the proposed square root matrix by itself:

$$\mathbb{M}^{1/2}\mathbb{M}^{1/2} = \mathbb{M}^{1/2}(\sqrt{\lambda_1} \mathbf{q}_1 \mathbf{q}_1^T + \ldots + \sqrt{\lambda_n} \mathbf{q}_n \mathbf{q}_n^T)$$

$$
\begin{aligned}
&= \sqrt{\lambda_1}\,\underbrace{\mathbb{M}^{1/2}\mathbf{q}_1}_{\sqrt{\lambda_1}\mathbf{q}_1}\,\mathbf{q}_1^T + \ldots + \sqrt{\lambda_n}\,\underbrace{\mathbb{M}^{1/2}\mathbf{q}_n}_{\sqrt{\lambda_n}\mathbf{q}_n}\,\mathbf{q}_n^T \\
&= \lambda_1\mathbf{q}_1\mathbf{q}_1^T + \ldots + \lambda_n\mathbf{q}_n\mathbf{q}_n^T \\
&= \mathbb{M}.
\end{aligned}
$$

♦

**Exercise 1.59** Let $\mathbb{M}$ be a symmetric and invertible real matrix. Show that $\mathbb{M}^{-1}$ is also a symmetric real matrix.

*Solution* Let $\lambda_1, \ldots, \lambda_n$ and $\mathbf{q}_1, \ldots, \mathbf{q}_n$ be eigenvalues and orthonormal eigenvectors of $\mathbb{M}$. Based on Exercise 1.24, we can deduce that $\mathbb{M}^{-1}$ has the spectral decomposition

$$
\mathbb{M}^{-1} = \frac{1}{\lambda_1}\mathbf{q}_1\mathbf{q}_1^T + \ldots + \frac{1}{\lambda_n}\mathbf{q}_n\mathbf{q}_n^T.
$$

Because $\mathbb{M}^{-1}$ is a linear combination of symmetric real matrices (see Exercise 1.54), it is clearly a symmetric real matrix as well. ♦

Using the spectral decomposition, we can also express the product of $\mathbb{M} \in \mathbb{R}^{n\times n}$ and $\mathbf{y}$ as

$$
\begin{aligned}
\mathbb{M}\mathbf{y} &= (\lambda_1\mathbf{q}_1\mathbf{q}_1^T + \ldots + \lambda_n\mathbf{q}_n\mathbf{q}_n^T)\mathbf{y} \\
&= \lambda_1\mathbf{q}_1\mathbf{q}_1^T\mathbf{y} + \ldots + \lambda_n\mathbf{q}_n\mathbf{q}_n^T\mathbf{y} \\
&= \begin{bmatrix} | & & | \\ \lambda_1\mathbf{q}_1 & \cdots & \lambda_n\mathbf{q}_n \\ | & & | \end{bmatrix} \begin{bmatrix} \mathbf{q}_1^T\mathbf{y} \\ \vdots \\ \mathbf{q}_n^T\mathbf{y} \end{bmatrix} \\
&= \begin{bmatrix} | & & | \\ \mathbf{q}_1 & \cdots & \mathbf{q}_n \\ | & & | \end{bmatrix} \begin{bmatrix} \lambda_1 & & \\ & \ddots & \\ & & \lambda_n \end{bmatrix} \begin{bmatrix} - & \mathbf{q}_1 & - \\ & \vdots & \\ - & \mathbf{q}_n & - \end{bmatrix} \mathbf{y}.
\end{aligned}
$$

Thus $\mathbb{M}$ can be written as $\mathbb{Q}\mathbb{\Lambda}\mathbb{Q}^T$ where $\mathbb{Q}$ is a matrix whose columns are orthonormal eigenvectors and $\mathbb{\Lambda}$ is a diagonal matrix of eigenvalues. We can interpret the actions of these three matrices in turn. The first matrix multiplication results in the vector of coordinates of $\mathbf{y}$ with respect to the eigenvector basis.

$$
\mathbb{Q}^T\mathbf{y} = \begin{bmatrix} - & \mathbf{q}_1 & - \\ & \vdots & \\ - & \mathbf{q}_n & - \end{bmatrix} \mathbf{y}
$$

$$= \begin{bmatrix} \langle \mathbf{y}, \mathbf{q}_1 \rangle \\ \vdots \\ \langle \mathbf{y}, \mathbf{q}_n \rangle \end{bmatrix}$$

Next, the $\mathbb{A}$ matrix multiplies each coordinate by the appropriate eigenvalue.

$$\begin{aligned} \mathbb{A}\mathbb{Q}^T \mathbf{y} &= \begin{bmatrix} \lambda_1 & & \\ & \ddots & \\ & & \lambda_n \end{bmatrix} \begin{bmatrix} \langle \mathbf{y}, \mathbf{q}_1 \rangle \\ \vdots \\ \langle \mathbf{y}, \mathbf{q}_n \rangle \end{bmatrix} \\ &= \begin{bmatrix} \lambda_1 \langle \mathbf{y}, \mathbf{q}_1 \rangle \\ \vdots \\ \lambda_n \langle \mathbf{y}, \mathbf{q}_n \rangle \end{bmatrix} \end{aligned}$$

Finally, multiplication by $\mathbb{Q}$ provides the linear combination of the eigenvector basis using those coordinates as the coefficients.

$$\begin{aligned} \mathbb{Q}\mathbb{A}\mathbb{Q}^T \mathbf{y} &= \begin{bmatrix} | & & | \\ \mathbf{q}_1 & \cdots & \mathbf{q}_n \\ | & & | \end{bmatrix} \begin{bmatrix} \lambda_1 \langle \mathbf{y}, \mathbf{q}_1 \rangle \\ \vdots \\ \lambda_n \langle \mathbf{y}, \mathbf{q}_n \rangle \end{bmatrix} \\ &= \lambda_1 \langle \mathbf{y}, \mathbf{q}_1 \rangle \begin{bmatrix} | \\ \mathbf{q}_1 \\ | \end{bmatrix} + \ldots + \lambda_n \langle \mathbf{y}, \mathbf{q}_n \rangle \begin{bmatrix} | \\ \mathbf{q}_n \\ | \end{bmatrix} \end{aligned}$$

This is exactly what we observed in Eq. 1.5.

**Exercise 1.60** Let $\mathbb{M}$ be a symmetric real matrix. Show that the trace of $\mathbb{M}$ equals the sum of its eigenvalues.[8]

*Solution* We will use the matrix form of spectral decomposition $\mathbb{M} = \mathbb{Q}\mathbb{A}\mathbb{Q}^T$ and the *cyclic permutation* property of trace (Exercise 1.51).

$$\begin{aligned} \operatorname{tr} \mathbb{M} &= \operatorname{tr} (\mathbb{Q}\mathbb{A}\mathbb{Q}^T) \\ &= \operatorname{tr} (\underbrace{\mathbb{Q}^T \mathbb{Q}}_{\mathbb{I}_n} \mathbb{A}) \\ &= \operatorname{tr} \mathbb{A} \end{aligned}$$

♦

[8] In fact, the trace of any square matrix equals the sum of its eigenvalues; symmetry is not required. To see why, you can consult Sec 3.8.4.2 of Gentle (2017).

While the spectral decomposition works for symmetric matrices, a closely related decomposition holds in more generality.

**Theorem 1.4 (Singular Value Theorem, Penney (2015) Sec 6.7)** *Let* $\mathbb{M}$ *be an* $n \times m$ *real matrix. There exists an orthonormal basis* $\mathbf{v}_1, \ldots, \mathbf{v}_m \in \mathbb{R}^m$ *for which* $\mathbb{M}\mathbf{v}_1, \ldots, \mathbb{M}\mathbf{v}_m \in \mathbb{R}^n$ *are orthogonal.*[9]

The singular value theorem implies that $\mathbb{M}$ can be represented by a *singular value decomposition*

$$\mathbb{M} = \sigma_1 \mathbf{u}_1 \mathbf{v}_1^T + \ldots + \sigma_d \mathbf{u}_d \mathbf{v}_d^T, \tag{1.6}$$

where $\sigma_1, \ldots, \sigma_d$ are non-negative real numbers (called *singular values* of $\mathbb{M}$), $\mathbf{u}_1, \ldots, \mathbf{u}_d$ are orthonormal vectors in $\mathbb{R}^n$, and $\mathbf{v}_1, \ldots, \mathbf{v}_d$ are orthonormal vectors in $\mathbb{R}^m$.

**Exercise 1.61** Let $\mathbb{M}$ be an $n \times m$ real matrix. How do you know that the number of terms in a singular value decomposition of $\mathbb{M}$ cannot be more than $\min(n, m)$.

*Solution* The vectors $\mathbf{u}_1, \ldots, \mathbf{u}_d \in \mathbb{R}^n$ are linearly independent, so there cannot be more than $n$ of them. Likewise, the vectors $\mathbf{v}_1, \ldots, \mathbf{v}_d \in \mathbb{R}^m$ are linearly independent, so there cannot be more than $m$ of them. ♦

As with spectral decompositions, there are also matrix product representations of the singular value decomposition (SVD). If $\mathbb{M} \in \mathbb{R}^{n\times m}$ and $d := \min(m, n)$, then

$$\mathbb{M} = \mathbb{U}\mathbb{S}\mathbb{V}^T,$$

where $\mathbb{U} \in \mathbb{R}^{n\times d}$ and $\mathbb{V} \in \mathbb{R}^{m\times d}$ both have orthonormal columns and $\mathbb{S} \in \mathbb{R}^{d\times d}$ is diagonal and has only non-negative values. Matching this to the original formulation in Eq. 1.6, $\mathbf{u}_1, \ldots, \mathbf{u}_d$ are the columns of $\mathbb{U}$, $\mathbf{v}_1, \ldots, \mathbf{v}_d$ are the columns of $\mathbb{V}$, and $\sigma_1, \ldots, \sigma_d$ are the entries along the diagonal of $\mathbb{S}$.

There is often some freedom in representing a matrix with an SVD. When the same singular value is repeated, the corresponding vectors in $\mathbb{U}$ and/or $\mathbb{V}$ can be replaced by certain other vectors in their span. When $\mathbb{M}$ is not full rank the square matrix $\mathbb{S}$ can be truncated to omit the singular values of 0; the corresponding columns of $\mathbb{U}$ and $\mathbb{V}$ are likewise removed to make the multiplication remain well-defined.[10] Conversely, columns can be added to $\mathbb{U}$ (and/or $\mathbb{V}$) in order to make the column vectors comprise an orthonormal basis; the same number of rows (and/or columns) of zeros must be added to $\mathbb{S}$.

We can understand the roles of the individual matrices in an SVD, analogously to spectral decomposition. First, $\mathbb{V}^T\mathbf{w}$ represents $\mathbf{w}$ as a vector of coordinates

[9]Note that some of $\mathbb{M}\mathbf{v}_1, \ldots, \mathbb{M}\mathbf{v}_m$ may be the zero vector.

[10]In that case, the columns of $\mathbb{U}$ are an orthonormal basis for the column space of $\mathbb{M}$, while the columns of $\mathbb{V}$ are an orthonormal basis for the row space. This representation makes it clear that the dimension of the column space is the same as the dimension of the row space.

with respect to $\{\mathbf{v}_1, \ldots, \mathbf{v}_d\}$. Next, multiplication by $\mathbb{S}$ rescales those coordinates according to the singular values. Finally, multiplication by $\mathbb{U}$ uses those rescaled coordinates as the coefficients of $\mathbf{u}_1, \ldots, \mathbf{u}_d$.

Let us think about the relationship between the two decompositions that we have introduced. The singular value theorem says that there is an orthonormal basis in the domain that maps to a rescaled version of an orthonormal basis in the range. The spectral theorem says that for symmetric matrices, these orthonormal bases can be chosen to be the exact same basis. A subtle difference is that, by convention, the basis vectors in an SVD are chosen to make the scaling values (diagonals of $\mathbb{S}$) non-negative, whereas a spectral decomposition can have negative scaling values (diagonals of $\mathbb{\Lambda}$).

**Exercise 1.62** Use a singular value decomposition for $\mathbb{M} \in \mathbb{R}^{n \times m}$ to find a spectral decomposition of $\mathbb{M}^T\mathbb{M}$.

*Solution* Writing $\mathbb{M} = \mathbb{U}\mathbb{S}\mathbb{V}^T$,

$$\begin{aligned} \mathbb{M}^T\mathbb{M} &= (\mathbb{U}\mathbb{S}\mathbb{V}^T)^T(\mathbb{U}\mathbb{S}\mathbb{V}^T) \\ &= \mathbb{V}\mathbb{S}^T \underbrace{\mathbb{U}^T\mathbb{U}}_{\mathbb{I}} \mathbb{S}\mathbb{V}^T \\ &= \mathbb{V}\mathbb{S}^2\mathbb{V}^T. \end{aligned}$$

By comparison to the matrix form of spectral decomposition, we see that $\mathbb{M}^T\mathbb{M}$ has eigenvalues equal to the squares of the singular values of $\mathbb{M}$, and the corresponding eigenvectors are the columns of $\mathbb{V}$. ♦

**Exercise 1.63** Explain why $\mathbb{M}^T\mathbb{M}$ is invertible if and only if the columns of $\mathbb{M} \in \mathbb{R}^{n \times m}$ are linearly independent.

## 1.13 Orthogonal Projection Matrices

In $\mathbb{R}^n$, the orthogonal projection operator takes the form of a real matrix (Exercise 1.5).

**Exercise 1.64** Let $\mathbb{M} \in \mathbb{R}^{n \times n}$. Prove that $\mathbb{M}$ is symmetric if and only if $\langle \mathbf{v}, \mathbb{M}\mathbf{w} \rangle = \langle \mathbb{M}\mathbf{v}, \mathbf{w} \rangle$ for every $\mathbf{v}, \mathbf{w} \in \mathbb{R}^n$.

**Exercise 1.65** Let $\mathbb{H}$ be a real matrix. Use Exercise 1.64 to show that if $\mathbb{H}$ is an orthogonal projection matrix, then it must be symmetric.

**Exercise 1.66** Provide a formula for the matrix that maps $\mathbf{y}$ to its orthogonal projection onto the span of the unit vector $\mathbf{u} \in \mathbb{R}^n$.

*Solution* The orthogonal projection of $\mathbf{y}$ onto the span of $\mathbf{u}$ is $\langle \mathbf{y}, \mathbf{u} \rangle \mathbf{u}$. By rewriting this as $\mathbf{u}\mathbf{u}^T\mathbf{y}$, we realize that the matrix $\mathbf{u}\mathbf{u}^T$ maps any vector to its orthogonal projection onto the span of $\mathbf{u}$. ♦

In light of Exercise 1.66, we can interpret a spectral decomposition as a linear combination of orthogonal projection matrices onto orthogonal one-dimensional subspaces.

Additionally, Exercises 1.46 and 1.47 reveal what spectral decompositions look like for orthogonal projection matrices. If $\mathbf{u}_1, \ldots, \mathbf{u}_n$ comprise an orthonormal basis with $\mathbf{u}_1, \ldots, \mathbf{u}_m$ spanning $\mathcal{S}$, then the orthogonal projection matrix onto $\mathcal{S}$ can be expressed as

$$(1)\mathbf{u}_1\mathbf{u}_1^T + \ldots + (1)\mathbf{u}_m\mathbf{u}_m^T + (0)\mathbf{u}_{m+1}\mathbf{u}_{m+1}^T + \ldots + (0)\mathbf{u}_n\mathbf{u}_n^T.$$

It is easy to see why this linear combination is the orthogonal projection matrix by analyzing its behavior on the basis: it maps $\mathbf{u}_1, \ldots, \mathbf{u}_m$ to themselves and maps the rest of the basis vectors to 0. As a linear combination of symmetric matrices, the orthogonal projection matrix must also be symmetric. In fact, *any* symmetric matrix whose only eigenvalues are 0 and 1 will have such a spectral decomposition and is therefore an orthogonal projection matrix onto its eigenspace corresponding to eigenvalue 1.

**Exercise 1.67** Show that the trace of an orthogonal projection matrix equals the dimension of the subspace that it projects onto.

*Solution* From Exercise 1.60, we know that the trace of $\mathbb{H}$ equals the sum of its eigenvalues $\lambda_1, \ldots, \lambda_n$. Furthermore, because it is an orthogonal projection matrix, we know that it yields the spectral decomposition

$$\mathbb{H} = (1)\mathbf{q}_1\mathbf{q}_1^T + \ldots + (1)\mathbf{q}_m\mathbf{q}_m^T + (0)\mathbf{q}_{m+1}\mathbf{q}_{m+1}^T + \ldots + (0)\mathbf{q}_n\mathbf{q}_n^T,$$

where $\mathbf{q}_1, \ldots, \mathbf{q}_m$ are in the subspace that $\mathbb{H}$ projects onto and the rest are necessarily orthogonal to it. We see $m$ terms with the eigenvalue 1 and the remaining terms with the eigenvalue 0, so their sum is $m$ which is the dimension of the subspace that $\mathbb{H}$ projects onto. ♦

**Exercise 1.68** Let $\mathbb{M}$ be a matrix. Explain why the rank of the orthogonal projection matrix onto $C(\mathbb{M})$ must be exactly the same as the rank of $\mathbb{M}$.

*Solution* The equality of ranks follows from the stronger observation that the orthogonal projection matrix must have the exact same column space as $\mathbb{M}$. Every vector in $C(\mathbb{M})$ gets mapped to itself by the orthogonal projection matrix, so its column space is at least as large as $C(\mathbb{M})$. However, the orthogonal projection of any vector onto $C(\mathbb{M})$ must by definition be in $C(\mathbb{M})$, so the orthogonal projection matrix cannot map any vector to a result outside of $C(\mathbb{M})$. ♦

In the coming chapters, we will frequently need to deal with the orthogonal projections of vectors onto the column space of a matrix. We will now derive a simple formula for the orthogonal projection matrix onto the column space of a real matrix. The key is finding the right coefficients for the columns.

**Exercise 1.69** Let $\mathbb{M} \in \mathbb{R}^{n\times m}$ and $\mathbf{y} \in \mathbb{R}^n$. Explain why the *Normal equation*

$$\mathbb{M}^T\mathbb{M}\hat{\mathbf{b}} = \mathbb{M}^T\mathbf{y}$$

is satisfied by the coefficient vector $\hat{\mathbf{b}} \in \mathbb{R}^m$ if and only if $\mathbb{M}\hat{\mathbf{b}}$ is the orthogonal projection of $\mathbf{y}$ onto $C(\mathbb{M})$.

*Solution* The orthogonal projection $\mathbb{M}\hat{\mathbf{b}}$ is the unique vector in $C(\mathbb{M})$ with the property that $\mathbf{y}-\mathbb{M}\hat{\mathbf{b}} \perp C(\mathbb{M})$. It is equivalent to check that $\mathbf{y}-\mathbb{M}\hat{\mathbf{b}}$ is orthogonal to every column $\mathbf{v}_1,\ldots,\mathbf{v}_m$ of $\mathbb{M}$. Equivalently the following quantity should be equal to the zero vector:

$$\begin{aligned}
\mathbb{M}^T(\mathbf{y}-\mathbb{M}\hat{\mathbf{b}}) &= \begin{bmatrix} - & \mathbf{v}_1 & - \\ & \vdots & \\ - & \mathbf{v}_m & - \end{bmatrix}(\mathbf{y}-\mathbb{M}\hat{\mathbf{b}}) \\
&= \begin{bmatrix} \mathbf{v}_1^T(\mathbf{y}-\mathbb{M}\hat{\mathbf{b}}) \\ \vdots \\ \mathbf{v}_m^T(\mathbf{y}-\mathbb{M}\hat{\mathbf{b}}) \end{bmatrix}.
\end{aligned}$$

Setting this vector $\mathbb{M}^T(\mathbf{y}-\mathbb{M}\hat{\mathbf{b}})$ equal to the zero vector results in the Normal equation. ♦

**Exercise 1.70** Suppose $\mathbb{M} \in \mathbb{R}^{n\times m}$ has linearly independent columns. Provide a formula for the coefficient vector $\hat{\mathbf{b}}$ for which $\mathbb{M}\hat{\mathbf{b}}$ equals the orthogonal projection of $\mathbf{y} \in \mathbb{R}^n$ onto $C(\mathbb{M})$.

*Solution* Because the columns are linearly independent, we know that $\mathbb{M}^T\mathbb{M}$ is invertible and thus the Normal equation

$$\mathbb{M}^T\mathbb{M}\hat{\mathbf{b}} = \mathbb{M}^T\mathbf{y}$$

is uniquely solved by $\hat{\mathbf{b}} = (\mathbb{M}^T\mathbb{M})^{-1}\mathbb{M}^T\mathbf{y}$. ♦

**Exercise 1.71** Suppose $\mathbb{M} \in \mathbb{R}^{n\times m}$ has linearly independent columns. Provide a formula for the orthogonal projection matrix onto $C(\mathbb{M})$.

*Solution* We have already derived in Exercise 1.70 a formula for the desired coefficient vector $\hat{\mathbf{b}} = (\mathbb{M}^T\mathbb{M})^{-1}\mathbb{M}^T\mathbf{y}$, so we simply plug this into $\mathbb{M}\hat{\mathbf{b}}$ to find the orthogonal projection of $\mathbf{y}$ onto $C(\mathbb{M})$.

$$\begin{aligned}
\hat{\mathbf{y}} &= \mathbb{M}\hat{\mathbf{b}} \\
&= \mathbb{M}(\mathbb{M}^T\mathbb{M})^{-1}\mathbb{M}^T\mathbf{y}
\end{aligned}$$

Therefore, we see that $\mathbf{y}$ is mapped to its orthogonal projection onto $C(\mathbb{M})$ by the matrix $\mathbb{M}(\mathbb{M}^T\mathbb{M})^{-1}\mathbb{M}^T$. ♦

If the columns of $\mathbb{M} \in \mathbb{R}^{n\times m}$ are not linearly independent, one can still construct the orthogonal projection matrix by using a spectral decomposition of $\mathbb{M}$. Simply replace all non-zero eigenvalues with 1 to get the orthogonal projection matrix. However, in such cases one may still want to derive *coefficients* of the columns of $\mathbb{M}$ that map to the orthogonal projection. Here's the key tool for this endeavor.

**Definition (Moore–Penrose Inverse)** Let $\mathbb{M} \in \mathbb{R}^{n\times m}$ have singular value decomposition

$$\mathbb{M} = \sigma_1 \mathbf{u}_1 \mathbf{v}_1^T + \ldots + \sigma_d \mathbf{u}_d \mathbf{v}_d^T$$

with positive singular values. Then

$$\mathbb{M}^- := \frac{1}{\sigma_1}\mathbf{v}_1 \mathbf{u}_1^T + \ldots + \frac{1}{\sigma_d}\mathbf{v}_d \mathbf{u}_d^T$$

is called the **Moore–Penrose inverse** of $\mathbb{M}$.

The Moore–Penrose inverse is well-defined and unique for any real matrix. From the definition, we can see a *symmetric* relationship between $\mathbb{M}$ and $\mathbb{M}^-$, that is, $\mathbb{M}$ is also the Moore–Penrose inverse of $\mathbb{M}^-$. Furthermore, the column space of $\mathbb{M}$ is the row space of $\mathbb{M}^-$ and vice versa.

**Exercise 1.72** Show that $\mathbb{M}^-\mathbb{M}$ equals the orthogonal projection matrix onto the row space of $\mathbb{M}$.

**Exercise 1.73** Explain why the Moore–Penrose inverse of an invertible matrix must be its inverse.

Now we can return to the question of finding a coefficient vector that produces the orthogonal projection onto $\mathbb{M}$. Consider the proposal $\hat{\mathbf{b}} = (\mathbb{M}^T\mathbb{M})^-\mathbb{M}^T\hat{\mathbf{y}}$. Does it satisfy the Normal equation?

From Exercise 1.62, we can infer that $\mathbb{M}$ and $\mathbb{M}^T\mathbb{M}$ have the same row space, so $\mathbb{M}^T\mathbf{y}$ is in the row space of $\mathbb{M}^T\mathbb{M}$. Because $\mathbb{M}^T\mathbb{M}$ is symmetric, its row space equals its column space and therefore equals the row space of $(\mathbb{M}^T\mathbb{M})^-$. Exercise 1.72 enable us to confirm the Normal equation:

$$\begin{aligned}\mathbb{M}^T\mathbb{M}\hat{\mathbf{b}} &= (\mathbb{M}^T\mathbb{M})(\mathbb{M}^T\mathbb{M})^-\mathbb{M}^T\hat{\mathbf{y}} \\ &= \mathbb{M}^T\hat{\mathbf{y}}.\end{aligned}$$

**Exercise 1.74** Of all coefficient vectors that satisfy the Normal equation, show that $\hat{\mathbf{b}} = (\mathbb{M}^T\mathbb{M})^-\mathbb{M}^T\hat{\mathbf{y}}$ has the smallest norm.

**Exercise 1.75** Show that $\mathbb{M}\mathbb{M}^-\mathbb{M} = \mathbb{M}$.

## 1.14 Quadratic Forms

**Definition (Quadratic Form)** Any function that can be expressed as $f(\mathbf{v}) = \mathbf{v}^T \mathbb{M}\mathbf{v}$, where $\mathbb{M}$ is a symmetric matrix, is called a **quadratic form**.

**Exercise 1.76** For a unit vector $\mathbf{u}$, express the quadratic form $\mathbf{u}^T \mathbb{M}\mathbf{u}$ as a weighted average of the eigenvalues of $\mathbb{M} \in \mathbb{R}^{n \times n}$.

*Solution* Let $\mathbf{q}_1, \ldots, \mathbf{q}_n$ be an orthonormal basis of eigenvectors for $\mathbb{M}$ with eigenvalues $\lambda_1, \ldots, \lambda_n$. We can represent $\mathbf{u}$ with respect to the eigenvector basis as $\langle \mathbf{u}, \mathbf{q}_1 \rangle \mathbf{q}_1 + \ldots + \langle \mathbf{u}, \mathbf{q}_n \rangle \mathbf{q}_n$.

$$\begin{aligned}
\mathbf{u}^T \mathbb{M}\mathbf{u} &= \mathbf{u}^T \mathbb{M}(\langle \mathbf{u}, \mathbf{q}_1 \rangle \mathbf{q}_1 + \ldots + \langle \mathbf{u}, \mathbf{q}_n \rangle \mathbf{q}_n) \\
&= \mathbf{u}^T (\langle \mathbf{u}, \mathbf{q}_1 \rangle \underbrace{\mathbb{M}\mathbf{q}_1}_{\lambda_1 \mathbf{q}_1} + \ldots + \langle \mathbf{u}, \mathbf{q}_n \rangle \underbrace{\mathbb{M}\mathbf{q}_n}_{\lambda_n \mathbf{q}_n}) \\
&= \langle \mathbf{u}, \mathbf{q}_1 \rangle^2 \lambda_1 + \ldots + \langle \mathbf{u}, \mathbf{q}_n \rangle^2 \lambda_n
\end{aligned}$$

$\langle \mathbf{u}, \mathbf{q}_1 \rangle, \ldots, \langle \mathbf{u}, \mathbf{q}_n \rangle$ provide the coordinates of $\mathbf{u}$ with respect to the basis $\mathbf{q}_1, \ldots, \mathbf{q}_n$. Because $\mathbf{u}$ is a unit vector, the sum of these squared coordinates has to be 1. Additionally, the squared coordinates are non-negative. Consequently, we have expressed $\mathbf{u}^T \mathbb{M}\mathbf{u}$ as a weighted average of the eigenvalues; the weights are the squared coordinates of $\mathbf{u}$ with respect to the eigenvector basis. ♦

**Exercise 1.77** Identify a unit vector $\mathbf{u}$ that maximizes the quadratic form $\mathbf{u}^T \mathbb{M}\mathbf{u}$.

*Solution* From Exercise 1.76, we know that the quadratic form equals a weighted average of the eigenvalues. This weighted average is maximized by placing all of the weight on the largest eigenvalue, that is, by letting $\mathbf{u}$ be a principal eigenvector. Such a choice of $\mathbf{u}$ makes $\mathbf{u}^T \mathbb{M}\mathbf{u}$ equal to the largest eigenvalue. ♦

**Definition (Positive Definite)** Let $\mathbb{M}$ be a symmetric real matrix. $\mathbb{M}$ is called **positive definite** if for every vector $\mathbf{v} \neq \mathbf{0}$, the quadratic form $\mathbf{v}^T \mathbb{M}\mathbf{v}$ is greater than zero. A weaker condition, **positive semi-definite**, only requires every quadratic form to be greater than or equal to zero.

**Exercise 1.78** Given any real matrix $\mathbb{M}$, show that $\mathbb{M}^T \mathbb{M}$ is positive semi-definite.

*Solution* Exercise 1.54 established that the matrix in question is symmetric. The quadratic form

$$\begin{aligned}
\mathbf{v}^T (\mathbb{M}^T \mathbb{M})\mathbf{v} &= (\mathbf{v}^T \mathbb{M}^T)(\mathbb{M}\mathbf{v}) \\
&= (\mathbb{M}\mathbf{v})^T (\mathbb{M}\mathbf{v})
\end{aligned}$$

equals the squared norm of the vector $\mathbb{M}\mathbf{v}$ which is non-negative. ♦

**Exercise 1.79** Let $\mathbb{M}$ be a symmetric real matrix. Show that $\mathbb{M}$ is positive semi-definite if and only if its eigenvalues are all non-negative.

*Solution* From our work in Exercise 1.77, we have seen how to express the quadratic form as a linear combination of the eigenvalues

$$\mathbf{v}^T \mathbb{M} \mathbf{v} = \langle \mathbf{v}, \mathbf{q}_1 \rangle^2 \lambda_1 + \ldots + \langle \mathbf{v}, \mathbf{q}_n \rangle^2 \lambda_n.$$

If every eigenvalue is at least zero, then every term in this sum is non-negative so the quadratic form must be non-negative. Conversely, if $\lambda_j$ is negative, then the quadratic form arising from $\mathbf{v} = \mathbf{q}_j$ is negative, as it equals $\lambda_j$. ♦

**Exercise 1.80** Let $\mathbb{H} \in \mathbb{R}^{n \times n}$ be an orthogonal projection matrix, and let $\mathbf{v} \in \mathbb{R}^n$. Show that the squared length of $\mathbb{H}\mathbf{v}$ equals the quadratic form $\mathbf{v}^T \mathbb{H} \mathbf{v}$.

*Solution* Because $\mathbb{H}$ is symmetric and idempotent,

$$\begin{aligned} ||\mathbb{H}\mathbf{v}||^2 &= (\mathbb{H}\mathbf{v})^T (\mathbb{H}\mathbf{v}) \\ &= \mathbf{v}^T \mathbb{H}^T \mathbb{H} \mathbf{v} \\ &= \mathbf{v}^T \mathbb{H} \mathbf{v}. \end{aligned}$$

♦

## 1.15 Principal Components

Consider vectors $\mathbf{x}_1, \ldots, \mathbf{x}_n \in \mathbb{R}^m$. Recall that the *coordinates* of these vectors with respect to a given unit vector $\mathbf{u}$ are $\langle \mathbf{x}_1, \mathbf{u} \rangle, \ldots, \langle \mathbf{x}_n, \mathbf{u} \rangle$. By collecting $\mathbf{x}_1, \ldots, \mathbf{x}_n$ as the rows of a matrix $\mathbb{X}$, the coordinates with respect to $\mathbf{u}$ are conveniently the entries of the vector $\mathbb{X}\mathbf{u}$.

$$\begin{aligned} \mathbb{X}\mathbf{u} &= \begin{bmatrix} - \mathbf{x}_1 - \\ \vdots \\ - \mathbf{x}_n - \end{bmatrix} \mathbf{u} \\ &= \begin{bmatrix} \mathbf{x}_1^T \mathbf{u} \\ \vdots \\ \mathbf{x}_n^T \mathbf{u} \end{bmatrix} \end{aligned}$$

Similarly, if $\mathbf{u}_1, \ldots, \mathbf{u}_m$ comprise an orthonormal basis, then the coordinates of $\mathbf{x}$ with respect to this basis are $\langle \mathbf{x}, \mathbf{u}_1 \rangle, \ldots, \langle \mathbf{x}, \mathbf{u}_m \rangle$. With the basis vectors as the columns of $\mathbb{U}$, the coordinates of $\mathbf{x}$ are the entries of the *row* vector $\mathbf{x}^T \mathbb{U}$.

$$\begin{aligned}\mathbf{x}^T\mathbb{U} &= \mathbf{x}^T\begin{bmatrix} | & & | \\ \mathbf{u}_1 & \cdots & \mathbf{u}_m \\ | & & | \end{bmatrix} \\ &= \begin{bmatrix}\mathbf{x}^T\mathbf{u}_1 & \cdots & \mathbf{x}^T\mathbf{u}_m\end{bmatrix}\end{aligned}$$

Let us put these ideas together. The coordinates of $\mathbf{x}_1, \ldots, \mathbf{x}_n$ with respect to $\mathbf{u}_1, \ldots, \mathbf{u}_m$ are contained in the matrix $\mathbb{XU}$.

$$\begin{aligned}\mathbb{XU} &= \begin{bmatrix} - & \mathbf{x}_1 & - \\ & \vdots & \\ - & \mathbf{x}_n & - \end{bmatrix}\begin{bmatrix} | & & | \\ \mathbf{u}_1 & \cdots & \mathbf{u}_m \\ | & & | \end{bmatrix} \\ &= \begin{bmatrix}\mathbf{x}_1^T\mathbf{u}_1 & \cdots & \mathbf{x}_1^T\mathbf{u}_m \\ \vdots & \ddots & \vdots \\ \mathbf{x}_n^T\mathbf{u}_1 & \cdots & \mathbf{x}_n^T\mathbf{u}_m\end{bmatrix}\end{aligned}$$

The $i$th row provides the coordinates of $\mathbf{x}_i$ with respect to the basis; the $j$th column provides the coordinates of the *data* vectors with respect to basis vector $\mathbf{u}_j$.

**Exercise 1.81** Let $\mathbf{x}_1, \ldots, \mathbf{x}_n$ be the rows of a real matrix $\mathbb{X}$. Show that the quadratic form $\mathbf{u}^T(\frac{1}{n}\mathbb{X}^T\mathbb{X})\mathbf{u}$ is equal to the average of the squares of the coefficients of $\mathbf{x}_1, \ldots, \mathbf{x}_n$ with respect to $\mathbf{u}$.

*Solution* We will first express the quadratic form in terms of the squared norm of a vector.

$$\begin{aligned}\mathbf{u}^T(\tfrac{1}{n}\mathbb{X}^T\mathbb{X})\mathbf{u} &= \tfrac{1}{n}(\mathbb{X}\mathbf{u})^T(\mathbb{X}\mathbf{u}) \\ &= \tfrac{1}{n}\|\mathbb{X}\mathbf{u}\|^2\end{aligned}$$

The entries of the vector $\mathbb{X}\mathbf{u}$ are the coefficients of $\mathbf{x}_1, \ldots, \mathbf{x}_n$ with respect to $\mathbf{u}$. Its squared norm is the sum of its squared entries, so $\frac{1}{n}\|\mathbb{X}\mathbf{u}\|^2$ is the average of the squared coefficients. ♦

**Exercise 1.82** Let $\mathbf{x}_1, \ldots, \mathbf{x}_n$ be the rows of the matrix $\mathbb{X}$. Show that $\frac{1}{n}\mathbb{X}^T\mathbb{X}$ is the matrix whose $(j, k)$-entry is the average of the product of the $j$th and $k$th coordinates of the vectors $\mathbf{x}_1, \ldots, \mathbf{x}_n$.

*Solution* The product of the matrices

$$\mathbb{X}^T\mathbb{X} = \begin{bmatrix} | & & | \\ \mathbf{x}_1 & \cdots & \mathbf{x}_n \\ | & & | \end{bmatrix}\begin{bmatrix} - & \mathbf{x}_1 & - \\ & \vdots & \\ - & \mathbf{x}_n & - \end{bmatrix}$$

has as its $(j, k)$-entry the inner product of the $j$th row of $\mathbb{X}^T$ and the $k$th column of $\mathbb{X}$. With $x_{i,j}$ denoting the $j$th coordinate of $\mathbf{x}_i$, this inner product equals $\sum_i x_{i,j}x_{i,k}$. When multiplied by $1/n$, this entry is indeed the average of the products of the coordinates. By thinking about summing over the observations, $\frac{1}{n}\mathbb{X}^T\mathbb{X}$ can also be understood as an average of rank-1 matrices $\frac{1}{n}\mathbf{x}_i\mathbf{x}_i^T$. ♦

**Exercise 1.83** Let $\mathbf{x}_1, \ldots, \mathbf{x}_n$ be the rows of a real matrix $\mathbb{X}$. Show that the average squared length $\frac{1}{n}\sum_i \|\mathbf{x}_i\|^2$ equals the sum of the eigenvalues of $\frac{1}{n}\mathbb{X}^T\mathbb{X}$.

*Solution* By Parseval's identity, the squared norm equals the sum of the squared coordinates using any basis; let us consider the orthonormal eigenvectors $\mathbf{q}_1, \ldots, \mathbf{q}_m$ of $\frac{1}{n}\mathbb{X}^T\mathbb{X}$, with $\lambda_1, \ldots, \lambda_m$ denoting their eigenvalues. Recall that Exercise 1.81 allows us to rewrite the average of squared coefficients as a quadratic form.

$$\begin{aligned}
\frac{1}{n}\sum_i \|\mathbf{x}_i\|^2 &= \frac{1}{n}\sum_i (\langle \mathbf{x}_i, \mathbf{q}_1\rangle^2 + \ldots + \langle \mathbf{x}_i, \mathbf{q}_m\rangle^2) \\
&= \frac{1}{n}\sum_i \langle \mathbf{x}_i, \mathbf{q}_1\rangle^2 + \ldots + \frac{1}{n}\sum_i \langle \mathbf{x}_i, \mathbf{q}_m\rangle^2 \\
&= \underbrace{\mathbf{q}_1^T(\tfrac{1}{n}\mathbb{X}^T\mathbb{X})\mathbf{q}_1}_{\lambda_1} + \ldots + \underbrace{\mathbf{q}_m^T(\tfrac{1}{n}\mathbb{X}^T\mathbb{X})\mathbf{q}_m}_{\lambda_m}
\end{aligned}$$

Exercise 1.76 demonstrated that a quadratic form evaluated at a unit eigenvector equals the corresponding eigenvalue. ♦

If the *zero vector* is (naively) used to *approximate* every one of $\mathbf{x}_1, \ldots, \mathbf{x}_n$, then the average squared approximation error is simply the average squared length $\frac{1}{n}\sum_i \|\mathbf{x}_i\|^2$. If instead $\mathbf{x}_1, \ldots, \mathbf{x}_n$ are approximated by $\langle \mathbf{x}_1, \mathbf{u}\rangle\mathbf{u}, \ldots, \langle \mathbf{x}_n, \mathbf{u}\rangle\mathbf{u}$, respectively, what would the average squared approximation error be?

For each $i$, we know that $\mathbf{x}_i - \langle \mathbf{x}_i, \mathbf{u}\rangle\mathbf{u}$ is orthogonal to $\mathbf{u}$. As in Exercises 1.37 and 1.38, the squared approximation error of $\mathbf{x}_i$ is reduced by $\langle \mathbf{x}_i, \mathbf{u}\rangle^2$. So the average squared approximation error is reduced by the average of the squared coefficients $\frac{1}{n}\sum_i \langle \mathbf{x}_i, \mathbf{u}\rangle^2$.

If we wanted to use a single basis vector to approximate the data vectors as well as possible *on average*, we would want to identify the basis vector with the largest average squared inner product with the data. Based on Exercise 1.81, we can express this quantity as a quadratic form, which we know how to maximize from Exercise 1.77.

$$\frac{1}{n}\sum_i \langle \mathbf{x}_i, \mathbf{u}\rangle^2 = \mathbf{u}^T(\tfrac{1}{n}\mathbb{X}^T\mathbb{X})\mathbf{u}$$

is maximized by any principal eigenvector of $\frac{1}{n}\mathbb{X}^T\mathbb{X}$.[11] Furthermore, the resulting reduction in average squared approximation error is exactly equal to that eigenvalue (Exercise 1.76). In fact, if the data are approximated with any of the eigenvectors of $\frac{1}{n}\mathbb{X}^T\mathbb{X}$, the resulting reduction in average squared approximation error equals the corresponding eigenvalue.

Suppose the eigenvalues of $\frac{1}{n}\mathbb{X}^T\mathbb{X}$ are ordered from largest to smallest: $\lambda_1 \geq \ldots \geq \lambda_m$ with orthonormal eigenvectors $\mathbf{q}_1, \ldots, \mathbf{q}_m$. To recap, approximating all of the data vectors with their best representatives from the span of only $\mathbf{q}_1$ reduces the average squared approximation error by $\lambda_1$. Bringing $\mathbf{q}_2$ into the mix, using the best representatives from the span of $\{\mathbf{q}_1, \mathbf{q}_2\}$, reduces the average squared approximation error by an additional $\lambda_2$, and so on. If the entire eigenvector basis is used, the reduction by $\sum_{i=1}^m \lambda_i$ brings the approximation error all the way down to zero. To understand why this makes sense, remember that $\mathbf{q}_1, \ldots, \mathbf{q}_m$ is a basis, so their span includes the data vectors.

The very same mathematics arise when one tries to find an approximation that "spreads out" a set of data points as much as possible.

**Definition (Empirical Mean)** The **empirical mean** of vectors $\mathbf{x}_1, \ldots, \mathbf{x}_n$ is $\frac{1}{n}\sum_i \mathbf{x}_i$.

**Definition (Empirical Variance)** The *empirical variance* of real numbers $a_1, \ldots, a_n$ is their average squared difference from their mean.

A number of basic facts about empirical means and empirical variances are special cases of more general results that we will see in Chap. 3; in particular, Sect. 3.2 clarifies the connection. Based on that section, one can understand more explicitly what the empirical mean looks like for vectors in $\mathbb{R}^m$: the $j$th coordinate of $\bar{\mathbf{x}} := \frac{1}{n}\sum_i \mathbf{x}_i$ is the average of the $j$th coordinates of vectors $\mathbf{x}_1, \ldots, \mathbf{x}_n$.

Suppose momentarily that the empirical mean (henceforth, simply called "mean") of $\mathbf{x}_1, \ldots, \mathbf{x}_n$ is the zero vector. Then the mean of their coordinates with respect to any unit vector $\mathbf{u}$ is zero (Exercise 3.7), so the empirical variance (henceforth, simply called "variance") of their coordinates is simply the sum of the squared coordinates. As before, we can express the sum of squared coordinates as the quadratic form

$$\frac{1}{n}\sum_i \langle \mathbf{x}_i, \mathbf{u}\rangle^2 = \mathbf{u}^T(\tfrac{1}{n}\mathbb{X}^T\mathbb{X})\mathbf{u}$$

which is maximized when $\mathbf{u}$ is a principal eigenvector of $\frac{1}{n}\mathbb{X}^T\mathbb{X}$. And again, the variance of the coordinates of the data vectors with respect to any eigenvector $\mathbf{q}_j$ is the eigenvalue $\lambda_j$. Among all vectors orthogonal to $\mathbf{q}_1$, the next eigenvector $\mathbf{q}_2$ produces coordinates of the data vectors with the largest possible variance, and so

[11] Of course, these eigenvectors are the same as those of $\mathbb{X}^T\mathbb{X}$, but the matrix with the $1/n$ factor has a more convenient interpretation of its eigenvalues, at least for people who like *averages* or *asymptotic stability*.

on. The sum of the variances of the coordinates with respect to any orthonormal basis, including the original standard basis, is equal to the sum of the eigenvalues.

Now we will drop the pretense that $\mathbf{x}_1, \ldots, \mathbf{x}_n$ have mean zero. Denote the *centered* data vectors by

$$\tilde{\mathbf{x}}_1 := \mathbf{x}_1 - \bar{\mathbf{x}}$$
$$\vdots$$
$$\tilde{\mathbf{x}}_n := \mathbf{x}_n - \bar{\mathbf{x}}$$

and the centered data matrix by

$$\widetilde{\mathbb{X}} = \begin{bmatrix} - & \tilde{\mathbf{x}}_1 & - \\ & \vdots & \\ - & \tilde{\mathbf{x}}_n & - \end{bmatrix}.$$

Based on Exercise 3.30, the variance of the coordinates of the data vectors with respect to a unit vector $\mathbf{u}$ is *exactly the same* as the variance of the coordinates of the *centered* data vectors. So the variances of the coordinates of $\mathbf{x}_1, \ldots, \mathbf{x}_n$ are determined by inspecting $\frac{1}{n}\widetilde{\mathbb{X}}^T\widetilde{\mathbb{X}}$. This matrix has a name, and it plays a key role in linear model theory.

**Definition (Empirical Covariance)** If $\mathbf{x} = (x_1, \ldots, x_m)$ and $\mathbf{y} = (y_1, \ldots, y_m)$ are vectors in $\mathbb{R}^m$ with means $\bar{x}$ and $\bar{y}$, their **empirical covariance**, denoted $\sigma_{\mathbf{x},\mathbf{y}}$ is $\frac{1}{n}\sum_i (x_i - \bar{x})(y_i - \bar{y})$. If $\mathbf{x}_1, \ldots, \mathbf{x}_n$ are vectors in $\mathbb{R}^m$, their **empirical covariance matrix** has $\sigma_{\mathbf{x}_i,\mathbf{x}_j}$ as its $(i, j)$ entry.

Let $\Sigma$ denote the empirical covariance matrix for $\mathbf{x}_1, \ldots, \mathbf{x}_n$ (henceforth, simply called "covariance matrix"). Exercise 1.82 reveals a convenient formula for covariance matrices:

$$\Sigma = \frac{1}{n}\widetilde{\mathbb{X}}^T\widetilde{\mathbb{X}}. \tag{1.7}$$

The unit vector for which the coordinates of $\mathbf{x}_1, \ldots, \mathbf{x}_n$ have the largest possible variance is precisely the unit vector for which the coordinates of $\tilde{\mathbf{x}}_1, \ldots, \tilde{\mathbf{x}}_n$ have the largest possible variance: the principal eigenvector of $\Sigma$; that variance is the largest eigenvalue of $\Sigma$. Among all vectors orthogonal to that principal eigenvector, the next eigenvector of $\Sigma$ produces coordinates of the data vectors with the largest possible variance, and so on. As before, the sum of the variances of the coordinates is equal to the sum of the eigenvalues of $\Sigma$.

The coordinates of the *centered* data vectors with respect to the covariance matrix's eigenvector basis ($\mathbf{q}_1, \ldots, \mathbf{q}_m$) are called the *principal components*. More specifically, with these coordinates arranged in a matrix

$$\widetilde{\mathbb{X}}\mathbb{Q} = \begin{bmatrix} - & \tilde{\mathbf{x}}_1 & - \\ & \vdots & \\ - & \tilde{\mathbf{x}}_n & - \end{bmatrix} \begin{bmatrix} | & & | \\ \mathbf{q}_1 & \cdots & \mathbf{q}_m \\ | & & | \end{bmatrix}$$

$$= \begin{bmatrix} \tilde{\mathbf{x}}_1^T \mathbf{q}_1 & \cdots & \tilde{\mathbf{x}}_1^T \mathbf{q}_m \\ \vdots & \ddots & \vdots \\ \tilde{\mathbf{x}}_n^T \mathbf{q}_1 & \cdots & \tilde{\mathbf{x}}_n^T \mathbf{q}_m \end{bmatrix},$$

the first column is the *first principal component* of the data, the second column is the *second principal component*, and so on.[12]

[12]Note that the principal components are not quite unique. For example, the eigenvector $-\mathbf{q}_1$ could have been chosen rather than $\mathbf{q}_1$, resulting in the negated version of the first principal component. In the case of a repeated eigenvalue, any orthonormal eigenvectors in the corresponding eigenspace could have been chosen.

# Chapter 2
# Least-Squares Linear Regression

Given pairs of numbers $(x_1, y_1), \ldots, (x_n, y_n)$ from a dataset, a scatterplot is a standard tool for visualizing the data and trying to determine a relationship between these $x$ and $y$-values. For instance, a "line of best fit" might be drawn on the picture to capture the relationship. In a scatterplot, each *observation* is drawn as a point. Alternatively, the *variables* $(x_1, \ldots, x_n)$ and $(y_1, \ldots, y_n)$ can be visualized as vectors in $\mathbb{R}^n$. A picture of this sort helps us figure out how to calculate a line of best fit and enables us to easily understand many of its properties. In this chapter, we will draw both of these types of pictures as we carefully study the *least-squares linear regression* method of fitting data.

## 2.1 Regressing Data

In the context of data analysis, a *variable* is a measurement that has been taken on a collection of objects.[1] An *observation* is an object that has had one or more measurements taken on it.[2] Often, each row of a dataset corresponds to an observation, while each column corresponds to a variable.

[1]Often these *observed variables* are mathematically modeled by *random variables*, which can cause some confusion. We will not make this connection until Chap. 4. For this chapter, we will only have data and no model.

[2]Do not take the word *measurement* too literally here. It does not necessarily involve a scientist reading numbers from a fancy instrument; it can be a person's age or sex, for example.



W. D. Brinda, *Visualizing Linear Models*,
https://doi.org/10.1007/978-3-030-64167-2_2

A variable is *quantitative* if it takes numerical values.[3] Alternatively, a variable is *categorical* if it has a limited set of possible values that are not numerical (e.g. sex or race). A categorical variable indicates group membership.[4]

It is often useful to designate some variables as *explanatory* and others as *response*.[5] If one of them happens last or is intuitively thought to be affected by the others, then it is a natural choice to be considered the response variable. Then your analysis might help you understand how the other variables affect it. Alternatively, if there is a variable that will not be measured on new observation in the future, you might prefer to call it the response variable. Then your analysis might help you predict its value based on the values of the other variables.

**Definition (Prediction Function)** A **prediction function** maps from the possible values of the explanatory variables to the possible values of the response variable.

**Definition (Residual)** Suppose a dataset has a *quantitative* response variable. Given a prediction function and a particular observation in the dataset, the **residual** of that observation is the actual value of the response variable minus the value of the prediction function evaluated at the observation's explanatory values.

**Definition (Regression)** Suppose a dataset has a *quantitative* response variable. **Regression** means using a dataset to select a prediction function that can be used to predict the response values of future observations from their explanatory values.[6]

Notice that the above definition says that regression "can be used" for prediction. Usually the point of regression is to predict future data or to understand the data-generating mechanism, but there are other possible motivations. In the vein of *descriptive statistics*, regression can be a way of *summarizing the relationship* among variables in a dataset. Or to completely break from statistical thinking, regression can also be used to *compress data*. In this chapter, we will not yet talk about future data or about data-generating mechanisms.

**Definition (Linear Regression)** In **linear regression**, the form of the prediction function is linear in its free parameters.

**Definition (Least-Squares Regression)** **Least-squares regression** means selecting a prediction function (among those under consideration) that *minimizes the sum of squared residuals*.

---

[3]The numbers should be meaningful as numbers rather than as codes for something else. For example, zip codes are not quantitative variables.

[4]Sometimes a categorical variable specifies groups that have a natural ordering (e.g. freshman, sophomore, junior, senior), in which case the variable is called *ordinal*.

[5]For simplicity, we will talk as if there is always exactly one response variable. That will indeed be the case throughout this text. The alternative context is called *multivariate* analysis.

[6]When the response variable is *categorical*, this selection is called *classification*.

In general, regression involves identifying the prediction function in a specified set that minimizes some objective function involving the data. *Linear* regression specifies a particular form for the set of prediction functions, while *least-squares* regression specifies a particular criterion to minimize. When used together (*least-squares linear* regression), an optimal prediction function is easily obtained because its parameter values are the coefficients of an orthogonal projection. The upcoming subsections will explain this, starting with the simplest case and gradually generalizing.

### 2.1.1 Least-Squares Location Regression

Before getting to scatterplots, let us study an even simpler situation. What if we have *zero* explanatory variables? Without explanatory variables to distinguish the observations, it is only natural to use the same single number to predict every response value. The prediction functions are simply constants; the residuals resulting from a given prediction function are the differences between the response values and that constant.

**Definition (Least-Squares Point)** If $y_1, \ldots, y_n \in \mathbb{R}$ are observations of a *response* variable, the **least-squares point**[7] is $\hat{a} \in \mathbb{R}$ that minimizes the sum of squared residuals when each $y_i$ is predicted by the same number,

$$\hat{a} := \operatorname*{argmin}_{a \in \mathbb{R}} \sum_i (y_i - a)^2.$$

Let us see if we can rewrite the above objective function in terms of vectors. The residuals are the entries of the vector

$$\begin{bmatrix} y_1 - a \\ \vdots \\ y_n - a \end{bmatrix} = \begin{bmatrix} y_1 \\ \vdots \\ y_n \end{bmatrix} - a \begin{bmatrix} 1 \\ \vdots \\ 1 \end{bmatrix}$$
$$= \mathbf{y} - a\mathbf{1}$$

with $\mathbf{y} = (y_1, \ldots, y_n)$ and with $\mathbf{1}$ denoting $(1, \ldots, 1) \in \mathbb{R}^n$. The sum of squared residuals equals the squared norm of the residual vector:

$$\|\mathbf{y} - a\mathbf{1}\|^2.$$

[7] The terms *least-squares point* and *location regression* are not common, but they are introduced in order to help clarify the coherence of this chapter's topics.

Thus, selecting a constant $a \in \mathbb{R}$ to predict every $y_1, \ldots, y_n$ is equivalent to selecting an $a$ for which $a\mathbf{1}$ predicts $\mathbf{y}$. The span of $\{\mathbf{1}\}$, i.e. the set $\{a\mathbf{1} : a \in \mathbb{R}\}$, can be called the *constant subspace* because each vector has the same value for all of its entries. Figure 2.1 compares a depiction of the response *observations* and a depiction of the response *variable* as a vector.

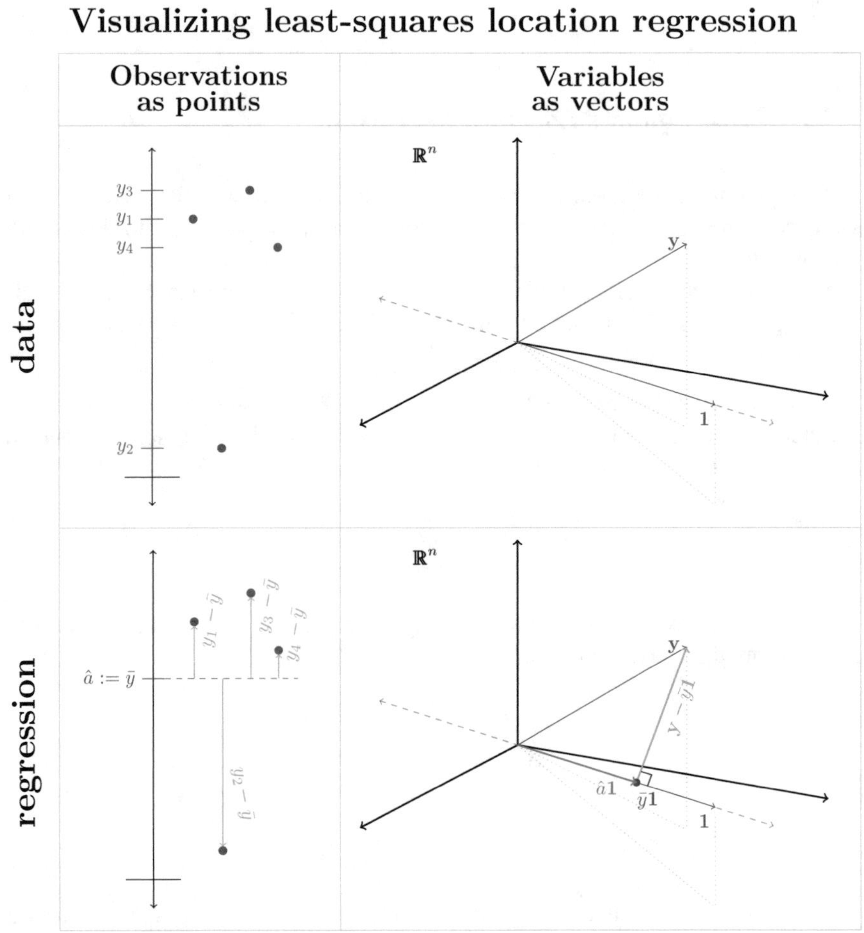


**Fig. 2.1** **Observations as points.** *data*: A one-dimensional scatterplot draws points with heights equal to the response values. (The points are spaced out horizontally only to help us see them better.) *regression*: Arrows represent residuals when the response values are all predicted by the height of the dotted line. The average of the response values is the least-squares point, the prediction that produces the smallest sum of squared residuals. **Variables as vectors.** *data*: A three-dimensional subspace is depicted that includes both the vectors $\mathbf{y}$ and $\mathbf{1}$. *regression*: Using the same value to predict every response is equivalent to using a vector in span$\{\mathbf{1}\}$ to predict $\mathbf{y}$. The sum of squared residuals equals the squared distance from $\mathbf{y}$ to its prediction vector, so the least-squares prediction vector is the orthogonal projection $\bar{y}\mathbf{1}$

**Exercise 2.1** Show that the entries of $\mathbf{v} = (v_1, \ldots, v_n)$ have mean zero if and only if $\mathbf{v}$ is orthogonal to $\mathbf{1} = (1, \ldots, 1)$.

*Solution* The average of the entries is proportional to the inner product of $\mathbf{v}$ with $\mathbf{1}$.

$$\frac{1}{n}\sum_i v_i = \frac{1}{n}\langle \mathbf{v}, \mathbf{1}\rangle$$

So the average is zero if and only if the inner product is zero. ♦

**Theorem 2.1 (Least-Squares Location Regression)** *Let $y_1, \ldots, y_n \in \mathbb{R}$. The objective function*

$$f(a) = \sum_i (y_i - a)^2$$

*is minimized by $\hat{a} := \bar{y}$, the average of the values.*

***Proof*** The vector formulation of the objective function $\|\mathbf{y} - a\mathbf{1}\|^2$ reveals that the set of possible prediction vectors is exactly the span of the unit vector $\mathbf{1}/\|\mathbf{1}\|$. The vector in that subspace that is closest to $\mathbf{y}$ is the orthogonal projection (Exercise 1.40), which we can calculate via Exercise 1.34

$$\begin{aligned}\hat{\mathbf{y}} &:= \frac{\langle \mathbf{y}, \mathbf{1}\rangle}{\|\mathbf{1}\|^2}\mathbf{1} \\ &= \underbrace{\frac{\sum_i y_i}{n}}_{\bar{y}}\mathbf{1}.\end{aligned}$$

Comparing this to $a\mathbf{1}$, we see that the minimizing coefficient must be $\bar{y}$. ■

**Exercise 2.2** Use the Pythagorean identity to decompose the average of the squared differences between the response values and $a \in \mathbb{R}$, that is $\frac{1}{n}\sum_i (y_i - a)^2$, into two terms, one of which is the empirical variance of $y_1, \ldots, y_n$.

*Solution* We can write $\sum_i (y_1 - a)^2$ as the squared norm $\|\mathbf{y} - a\mathbf{1}\|^2$. The vector $\mathbf{y} - a\mathbf{1}$ is the hypotenuse of the right triangle whose other two sides are $\mathbf{y} - \bar{y}\mathbf{1}$ and $\bar{y}\mathbf{1} - a\mathbf{1}$. By the Pythagorean identity,

$$\begin{aligned}\frac{1}{n}\sum_i (y_i - a)^2 &= \frac{1}{n}\|\mathbf{y} - a\mathbf{1}\|^2 \\ &= \frac{1}{n}[\|\mathbf{y} - \bar{y}\mathbf{1}\|^2 + \|\bar{y}\mathbf{1} - a\mathbf{1}\|^2] \\ &= \frac{1}{n}\left[\sum_i (y_i - \bar{y})^2 + n(\bar{y} - a)^2\right] \\ &= \frac{1}{n}\sum_i (y_i - \bar{y})^2 + (\bar{y} - a)^2.\end{aligned}$$

♦

While the *observations picture* is perfectly intuitive, it takes time to get used to the *variables picture*. If there were only 3 observations, it would be easy to envision or to draw the relevant vectors ($\mathbf{y}$ and $\mathbf{1}$ in this case) in $\mathbb{R}^3$. In general, though, the vectors inhabit $\mathbb{R}^n$, where $n$ is the number of observations. What if $n$ is larger than 3 or is left unspecified? We can still draw essentially the same picture, realizing that we are only depicting a three-dimensional subspace.[8] As long as no more than three vectors are being depicted, we know that they will all lie in a three-dimensional (at most) subspace of $\mathbb{R}^n$ and can therefore be accurately depicted in such a sketch.

## 2.1.2 Simple Linear Regression

**Definition (Least-Squares Line)** If $(x_1, y_1), \ldots, (x_n, y_n) \in \mathbb{R}^2$ are observations of an *explanatory* and a *response* variable, the **least-squares line** has the equation $y = \hat{a} + \hat{b}(x - \bar{x})$ where $\bar{x}$ is the average of $x_1, \ldots, x_n$, and the coefficients $\hat{a}, \hat{b} \in \mathbb{R}$ minimize the sum of squared residuals when each $y_i$ is predicted by a line,

$$(\hat{a}, \hat{b}) := \operatorname*{argmin}_{(a,b) \in \mathbb{R}^2} \sum_i (y_i - [a + b(x_i - \bar{x})])^2.$$

When the prediction function is a line of this form, the residuals are the entries of

$$\begin{bmatrix} y_1 - [a + b(x_1 - \bar{x})] \\ \vdots \\ y_n - [a + b(x_n - \bar{x})] \end{bmatrix} = \begin{bmatrix} y_1 \\ \vdots \\ y_n \end{bmatrix} - \left[ a \begin{bmatrix} 1 \\ \vdots \\ 1 \end{bmatrix} + b \left( \begin{bmatrix} x_1 \\ \vdots \\ x_n \end{bmatrix} - \bar{x} \begin{bmatrix} 1 \\ \vdots \\ 1 \end{bmatrix} \right) \right]$$
$$= \mathbf{y} - [a\mathbf{1} + b(\mathbf{x} - \bar{x}\mathbf{1})]$$

with $\mathbf{y} = (y_1, \ldots, y_n)$ and $\mathbf{x} = (x_1, \ldots, x_n)$. The sum of squared residuals equals the squared norm of the residual vector:

$$\|\mathbf{y} - [a\mathbf{1} + b(\mathbf{x} - \bar{x}\mathbf{1})]\|^2.$$

Given any prediction function, let us call its vector of predicted values a *prediction vector*. As in Sect. 2.1.1, the optimal coefficients are those that make the prediction vector equal to the orthogonal projection of $\mathbf{y}$ onto the subspace of possible prediction vectors. Figure 2.2 compares the observations picture and the variables picture in this scenario.

[8] If $n$ is left unspecified, do we have to worry about whether the drawings remain valid with $n < 3$? Not to worry, the drawing remains valid as $\mathbb{R}^n$ can be a subspace embedded within the three-dimensional picture.

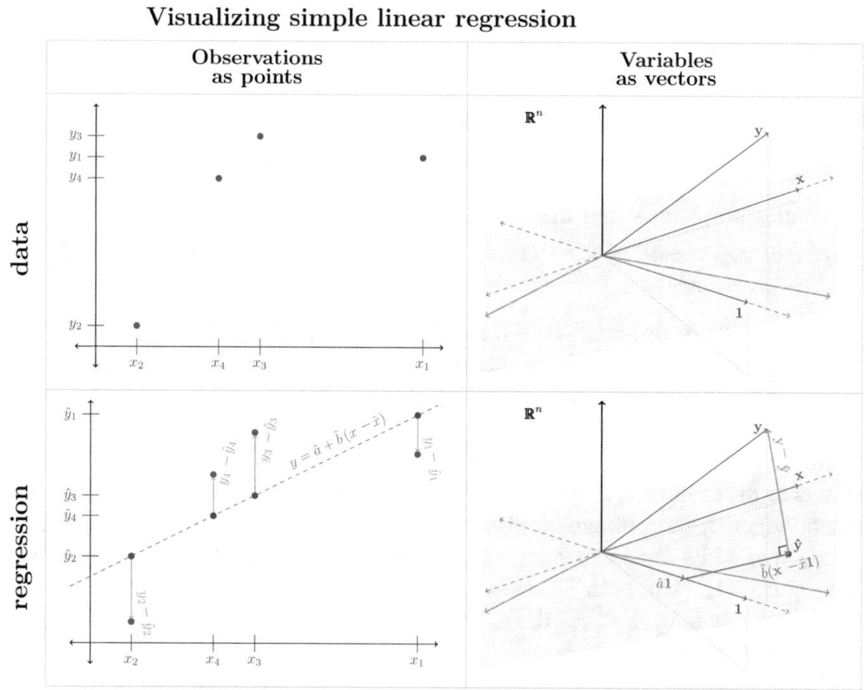


**Fig. 2.2** **Observations as points.** *data*: A scatterplot draws points with heights equal to their response values and horizontal positions corresponding to their explanatory values. *regression*: Arrows represent residuals when the response values are predicted by the height of the dotted line. Depicted is the least-squares line, the one that minimizes the sum of squared residuals. **Variables as vectors.** *data*: A three-dimensional subspace is depicted that includes the vectors $\mathbf{y}$, $\mathbf{1}$, and $\mathbf{x}$. *regression*: Using a line to predict the responses is equivalent to using a vector in span$\{\mathbf{1}, \mathbf{x}\}$ to predict $\mathbf{y}$. The sum of squared residuals equals the squared distance from $\mathbf{y}$ to its prediction vector, so the least-squares prediction vector is the orthogonal projection $\hat{\mathbf{y}}$. The coefficients of $\mathbf{1}$ and $\mathbf{x}-\bar{x}\mathbf{1}$ that lead to $\hat{\mathbf{y}}$ are the same as the corresponding coefficients of the least-squares line

Before endeavoring to work out the optimal coefficients, let us take a look at how neatly empirical variances and covariances can be expressed in terms of vectors. The empirical variance of $\mathbf{x}$ is

$$\begin{aligned}\sigma_{\mathbf{x}}^2 &:= \tfrac{1}{n}\sum_i (x_i - \bar{x})^2 \\ &= \tfrac{1}{n}\|\mathbf{x} - \bar{x}\mathbf{1}\|^2.\end{aligned}$$

The empirical covariance of $\mathbf{x}$ and $\mathbf{y}$ is

$$\begin{aligned}\sigma_{\mathbf{x},\mathbf{y}} &:= \tfrac{1}{n}\sum_i (x_i - \bar{x})(y_i - \bar{y}) \\ &= \tfrac{1}{n}\langle \mathbf{x} - \bar{x}\mathbf{1}, \mathbf{y} - \bar{y}\mathbf{1}\rangle.\end{aligned}$$

**Theorem 2.2 (Simple Linear Regression)** *Let* $(x_1, y_1), \ldots, (x_n, y_n) \in \mathbb{R}^2$. *The objective function*

$$f(a, b) = \sum_i (y_i - [a + b(x_i - \bar{x})])^2$$

*is minimized when its first argument is* $\hat{a} := \bar{y}$ *and its second is* $\hat{b} := \frac{\sigma_{\mathbf{x},\mathbf{y}}}{\sigma_{\mathbf{x}}^2}$, *the empirical covariance of* $\mathbf{x} = (x_1, \ldots, x_n)$ *and* $\mathbf{y} = (y_1, \ldots, y_n)$ *divided by the empirical variance of* $\mathbf{x}$.

***Proof*** The vector formulation of the objective function $\|\mathbf{y} - [a\mathbf{1} + b(\mathbf{x} - \bar{x}\mathbf{1})]\|^2$ reveals that the set of possible prediction vectors is exactly the span of $\mathbf{1}$ and $\mathbf{x} - \bar{x}\mathbf{1}$ (which is exactly the same as the span of $\mathbf{1}$ and $\mathbf{x}$ based on Exercise 1.4). The vector in that subspace that is closest to $\mathbf{y}$ is its orthogonal projection.

We can infer from our results in Sect. 2.1.1 that $\bar{x}\mathbf{1}$ is the orthogonal projection of $\mathbf{x}$ onto the span of $\mathbf{1}$, and so $\mathbf{x} - \bar{x}\mathbf{1}$ is orthogonal to $\mathbf{1}$. Because the two vectors are orthogonal, the orthogonal projection of $\mathbf{y}$ onto their span will be the same as the sum of its separate orthogonal projections onto span$\{\mathbf{1}\}$ and span$\{\mathbf{x} - \bar{x}\mathbf{1}\}$ (Exercise 1.41). First, as in Theorem 2.1 the orthogonal projection of $\mathbf{y}$ onto span$\{\mathbf{1}\}$ is $\bar{y}\mathbf{1}$. Next, as long as $\sigma_{\mathbf{x}}^2 > 0$, i.e. not all of $(x_1, \ldots, x_n)$ are the same value, the projection of $\mathbf{y}$ onto $\mathbf{x} - \bar{x}\mathbf{1}$ is

$$\frac{\langle \mathbf{y}, \mathbf{x} - \bar{x}\mathbf{1} \rangle}{\|\mathbf{x} - \bar{x}\mathbf{1}\|^2}(\mathbf{x} - \bar{x}\mathbf{1})$$

by the formula from Exercise 1.34. The coefficient can be rewritten as

$$\begin{aligned}\frac{\langle \mathbf{y}, \mathbf{x} - \bar{x}\mathbf{1} \rangle}{\|\mathbf{x} - \bar{x}\mathbf{1}\|^2} &= \frac{(1/n)\langle \mathbf{y} - \bar{y}\mathbf{1}, \mathbf{x} - \bar{x}\mathbf{1} \rangle}{(1/n)\|\mathbf{x} - \bar{x}\mathbf{1}\|^2} \\ &= \frac{\sigma_{\mathbf{x},\mathbf{y}}}{\sigma_{\mathbf{x}}^2}.\end{aligned}$$

(Verify for yourself that $\langle \mathbf{y} - \bar{y}\mathbf{1}, \mathbf{x} - \bar{x}\mathbf{1} \rangle$ is the same as $\langle \mathbf{y}, \mathbf{x} - \bar{x}\mathbf{1} \rangle$ or peek ahead to Exercise 3.29.) Finally, putting these together we obtain the orthogonal projection of $\mathbf{y}$ onto span$\{\mathbf{1}, \mathbf{x}\}$

$$\hat{\mathbf{y}} := \bar{y}\mathbf{1} + \frac{\sigma_{\mathbf{x},\mathbf{y}}}{\sigma_{\mathbf{x}}^2}(\mathbf{x} - \bar{x}\mathbf{1}).$$

Notice that if $\sigma_{\mathbf{x}}^2 = 0$, then $x_i - \bar{x} = 0$ for every observation. In that case, any value of $\hat{b}$ will work, but we will use the convention that $\hat{b} = 0$ so that it fits into a more general solution (Theorem 2.3) that we will develop shortly. ∎

**Exercise 2.3** Is it possible for the *least-squares line*'s sum of squared residuals to be greater than the *least-squares point*'s sum of squared residuals?

*Solution* The set of possible prediction functions corresponding to lines $\{f(x) = a + bx : a, b \in \mathbb{R}\}$ is strictly larger than the set of possible prediction functions corresponding to points $\{f(x) = a : a \in \mathbb{R}\}$. A line predicts every response value by the same number if its slope is zero. By definition, the least-squares line will use a slope of zero if and only if that leads to the smallest possible sum of squared residuals, in which case its sum of squared residuals would be equal to that of the least-squares point. ♦

**Exercise 2.4** The variables picture provides us with a more specific answer to the question posed in Exercise 2.3. Use the Pythagorean identity to quantify the difference between the least-squares point's sum of squared residuals and the least-squares line's sum of squared residuals.

*Solution* Because $\bar{y}\mathbf{1}$ is in the span of $\mathbf{1}$ and $\mathbf{x}$, we see that the least-squares line's residual vector $\mathbf{y}-\hat{\mathbf{y}}$ must be orthogonal to $\hat{\mathbf{y}}-\bar{y}\mathbf{1}$. Invoking the Pythagorean identity,

$$\|\mathbf{y} - \bar{y}\mathbf{1}\|^2 = \|\hat{\mathbf{y}} - \bar{y}\mathbf{1}\|^2 + \|\mathbf{y} - \hat{\mathbf{y}}\|^2.$$

The least-squares point's sum of squared residuals is larger than the least-squares line's sum of squared residuals by $\|\hat{\mathbf{y}} - \bar{y}\mathbf{1}\|^2$. ♦

Rather than the empirical variances and covariance, people sometimes prefer the *unbiased estimators*[9] which use $\frac{1}{n-1}$ rather than $\frac{1}{n}$. If all empirical variance and covariances in the simple linear regression slope are replaced by their unbiased estimator counterparts, the formula still works because the factor in question cancels out anyway.

The least-squares line can be expressed in the usual slope-intercept form[10] as $y = (\hat{a} - \hat{b}\bar{x}) + \hat{b}x$.

Let us take a look at one more formulation of the least-squares line, this time in terms of the "standardized" variables. To *standardize* a vector, you subtract its mean from each entry then divide every entry by the standard deviation. We will see that the least-squares line for the standardized variables has intercept 0 and slope equal to the following statistic.

**Definition (Correlation)** The **correlation** between $\mathbf{x} = (x_1, \ldots, x_n)$ and $\mathbf{y} = (y_1, \ldots, y_n)$ is

[9] More precisely, these statistics are unbiased estimates of the variances and covariance between the random variables $X$ and $Y$, assuming $(x_1, y_1), \ldots, (x_n, y_n)$ were iid draws from the distribution of $(X, Y)$ see Exercise 3.36.

[10] Perhaps you would rather use the slope-intercept parameterization $y = a + bx$ in the first place rather than subtracting $\bar{x}$ from $x$. I have concluded that there is very good reason to subtract $\bar{x}$, so bear with me. Either way you parameterize it, you will derive the same least-squares line.

$$\rho_{\mathbf{x},\mathbf{y}} := \frac{\sigma_{\mathbf{x},\mathbf{y}}}{\sigma_{\mathbf{x}}\sigma_{\mathbf{y}}}.$$

**Exercise 2.5** Show that the correlation between two vectors equals the empirical covariance of their standardized versions.

Substituting $\rho_{\mathbf{x},\mathbf{y}}\frac{\sigma_{\mathbf{y}}}{\sigma_{\mathbf{x}}}$ for $\frac{\sigma_{\mathbf{x},\mathbf{y}}}{\sigma_{\mathbf{x}}^2}$ in the least-squares line and rearranging,

$$\frac{y - \bar{y}}{\sigma_{\mathbf{y}}} = \rho_{\mathbf{x},\mathbf{y}}\frac{x - \bar{x}}{\sigma_{\mathbf{x}}}.$$

In other words, if the explanatory variable is $z$ standard deviations above its mean, then the least-squares line predicts the response variable to be $\rho_{\mathbf{x},\mathbf{y}}z$ standard deviations above its mean. Thus the response variable is predicted to be *less extreme* than the explanatory variable. Least-squares fitting was pioneered in the late nineteenth century by the eminent British intellectual Sir Francis Galton who first demonstrated it to predict men's heights using their fathers' heights. The men's heights were indeed found to be less extreme on average than their fathers' heights, a phenomenon called by Galton "regression toward mediocrity";[11] the term *regression* caught on and is now used broadly for fitting quantitative response data.

## 2.1.3 *Multiple Linear Regression*

To make sure you understand how the pictures continue to generalize, we will now move on to the case in which the data comprises a quantitative response variable and *two* quantitative explanatory variables. The ordinary scatterplot does not have enough dimensions to let us visualize these points. We require a two-dimensional real plane just to index all the possible pairs of explanatory variable values. We need a third axis rising perpendicularly from this plane to represent the possible response variable values. This picture is called a *3-D scatterplot*.

Let $x_1^{(1)}, \ldots, x_n^{(1)}$ and $x_1^{(2)}, \ldots, x_n^{(2)}$ be the values of two quantitative explanatory variables with $\bar{x}^{(1)}$ and $\bar{x}^{(2)}$ as their averages. Now we consider regression functions of the form $f_{a,b_1,b_2}(x^{(1)}, x^{(2)}) = a + b_1(x^{(1)} - \bar{x}^{(1)}) + b_2(x^{(2)} - \bar{x}^{(2)})$ with $(a, b_1, b_2)$ ranging over $\mathbb{R}^3$. Each possible $(a, b_1, b_2)$ defines a different *plane*. Each plane creates a residual at each observation, the vertical difference between the point and the height of the plane at the location of the two explanatory variable values. One of those planes has the smallest possible sum of squared residuals.

A *hyperplane* generalizes the concept of line and plane to arbitrarily many dimensions. With $m$ explanatory variables, one can imagine the observations as

[11] Of course, individuals also have a chance to be more exceptional than their parents in any given characteristic, and as a result aggregate population characteristics remain relatively stable.

points in $\mathbb{R}^{m+1}$ and seek the $m$-dimensional hyperplane that minimizes the sum of squared residuals.

**Definition (Least-Squares Hyperplane)** If $(x_1^{(1)}, \ldots, x_1^{(m)}, y_1), \ldots, (x_n^{(1)}, \ldots, x_n^{(m)}, y_n) \in \mathbb{R}^{m+1}$ are observations of $m$ *explanatory* and one *response* variable, the **least-squares hyperplane** has the equation

$$y = \hat{a} + \hat{b}_1(x^{(1)} - \bar{x}^{(1)}) + \ldots + \hat{b}_m(x^{(m)} - \bar{x}^{(m)}),$$

where $\bar{x}^{(j)}$ is the average of $x_1^{(j)}, \ldots, x_n^{(j)}$, and the coefficients $\hat{a}, \hat{b}_1, \ldots, \hat{b}_m \in \mathbb{R}$ minimize the sum of squared residuals when each $y_i$ is predicted by a hyperplane,

$$(\hat{a}, \hat{b}_1, \ldots, \hat{b}_m) := \underset{(a,b_1,\ldots,b_m)\in\mathbb{R}^{m+1}}{\operatorname{argmin}} \sum_i (y_i - [a + b_1(x_i^{(1)} - \bar{x}^{(1)}) + \ldots + b_m(x_i^{(m)} - \bar{x}^{(m)})])^2.$$

When the prediction function is a hyperplane of the above form, the residuals are the entries of

$$\begin{bmatrix} y_1 - [a + b_1(x_1^{(1)} - \bar{x}^{(1)}) + \ldots + b_m(x_1^{(m)} - \bar{x}^{(m)})] \\ \vdots \\ y_n - [a + b_1(x_n^{(1)} - \bar{x}^{(1)}) + \ldots + b_m(x_n^{(m)} - \bar{x}^{(m)})] \end{bmatrix}$$

$$= \begin{bmatrix} y_1 \\ \vdots \\ y_n \end{bmatrix} - \left[ a \begin{bmatrix} 1 \\ \vdots \\ 1 \end{bmatrix} + b_1 \begin{bmatrix} x_1^{(1)} - \bar{x}^{(1)} \\ \vdots \\ x_n^{(1)} - \bar{x}^{(1)} \end{bmatrix} + \ldots + b_m \begin{bmatrix} x_1^{(m)} - \bar{x}^{(m)} \\ \vdots \\ x_n^{(m)} - \bar{x}^{(m)} \end{bmatrix} \right]$$

$$= \mathbf{y} - [a\mathbf{1} + \widetilde{\mathbb{X}}\mathbf{b}]$$

with $\mathbf{y} = (y_1, \ldots, y_n)$, $\mathbf{b} = (b_1, \ldots, b_m)$ and $\widetilde{\mathbb{X}}$ representing the centered explanatory data matrix

$$\widetilde{\mathbb{X}} := \begin{bmatrix} x_1^{(1)} - \bar{x}^{(1)} & \cdots & x_1^{(m)} - \bar{x}^{(m)} \\ \vdots & \ddots & \vdots \\ x_n^{(1)} - \bar{x}^{(1)} & \cdots & x_n^{(m)} - \bar{x}^{(m)} \end{bmatrix}$$

which played a role in Sect. 1.15.[12] The sum of squared residuals equals the squared norm of the residual vector:

[12] In that section, the centered data matrix was defined by subtracting the empirical mean vector from each row of the data matrix. Convince yourself that the expression given here is equivalent.

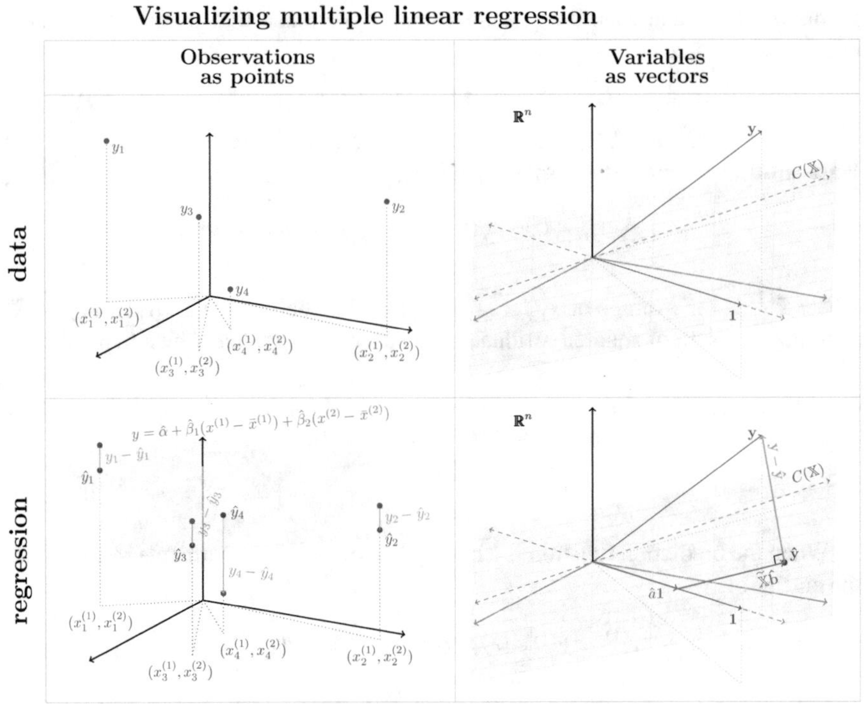


**Fig. 2.3** **Observations as points.** *data*: A three-dimensional scatterplot draws points with heights equal to their response values and positions in the other two directions corresponding to their two explanatory values. *regression*: Arrows represent residuals when the response values are predicted by the height of the least-squares plane, the plane that minimizes the sum of squared residuals. **Variables as vectors.** *data*: A three-dimensional subspace is depicted that includes $\mathbf{y}$, $\mathbf{1}$, and a one-dimensional subspace of $C(\mathbb{X})$. *regression*: Using a hyperplane to predict the responses is equivalent to using a vector in span$\{\mathbf{1}, \mathbf{x}^{(1)}, \ldots, \mathbf{x}^{(m)}\}$ to predict $\mathbf{y}$. The sum of squared residuals equals the squared distance from $\mathbf{y}$ to its prediction vector, so the least-squares prediction vector is the orthogonal projection $\hat{\mathbf{y}}$. The coefficients that lead to $\hat{\mathbf{y}}$ are the same as the corresponding coefficients of the least-squares hyperplane

$$\|\mathbf{y} - (a\mathbf{1} + \widetilde{\mathbb{X}}\mathbf{b})\|^2.$$

Figure 2.3 demonstrates the observations picture and the variables picture in this context.

**Theorem 2.3 (Multiple Linear Regression)** *Let* $(x_1^{(1)}, \ldots, x_1^{(m)}, y_1), \ldots, (x_n^{(1)}, \ldots, x_n^{(m)}, y_n) \in \mathbb{R}^{m+1}$, *and let* $\bar{x}^{(j)}$ *represent the average of* $x_1^{(j)}, \ldots, x_n^{(j)}$. *The objective function*

$$f(a, b_1, \ldots, b_m) = \sum_i (y_i - [a + b_1(x_i^{(1)} - \bar{x}^{(1)}) + \ldots + b_m(x_i^{(m)} - \bar{x}^{(m)})])^2$$

*is minimized when its first argument is* $\hat{a} := \bar{y}$ *and the remaining arguments are*

$$(\hat{b}_1, \ldots, \hat{b}_m) := \Sigma^- \begin{bmatrix} \sigma_{\mathbf{x}^{(1)},\mathbf{y}} \\ \vdots \\ \sigma_{\mathbf{x}^{(m)},\mathbf{y}}, \end{bmatrix}$$

*where* $\Sigma$ *is the explanatory data's empirical covariance matrix and* $\sigma_{\mathbf{x}^{(j)},\mathbf{y}}$ *is the empirical covariance between* $\mathbf{x}^{(j)} = (x_1^{(j)}, \ldots, x_n^{(j)})$ *and* $\mathbf{y} = (y_1, \ldots, y_n)$.

***Proof*** Let $\mathbb{X}$ represent the original explanatory data matrix, having $\mathbf{x}^{(j)} := (x_1^{(j)}, \ldots, x_n^{(j)})$ as its $j$th column, for each $j$. Let $\widetilde{\mathbb{X}}$ represent the centered version.

The vector formulation of the objective function $\|\mathbf{y} - [a\mathbf{1} + \widetilde{\mathbb{X}}\mathbf{b}]\|^2$ reveals that the set of possible prediction vectors is exactly the span of $\mathbf{1}$ and the columns of $\widetilde{\mathbb{X}}$ (which is exactly the same as the span of $\mathbf{1}$ and the columns of the original data matrix $\mathbb{X}$ based on Exercise 1.4). The vector in that subspace that is closest to $\mathbf{y}$ is the orthogonal projection of $\mathbf{y}$ onto the subspace.

We can infer from our results in Sect. 2.1.1 that $\bar{x}^{(j)}\mathbf{1}$ is the orthogonal projection of $\mathbf{x}_j$ onto the span of $\mathbf{1}$, and so the $j$th column of $\widetilde{\mathbb{X}}$, which is $\mathbf{x}^{(j)} - \bar{x}^{(j)}\mathbf{1}$, is orthogonal to $\mathbf{1}$. In other words, every column of $\widetilde{\mathbb{X}}$ is orthogonal to $\mathbf{1}$. Because span$\{\mathbf{1}\}$ and $C(\widetilde{\mathbb{X}})$ are orthogonal subspaces, Exercise 1.41 implies that the orthogonal projection of $\mathbf{y}$ onto their span will be the same as the sum of its separate orthogonal projections onto span$\{\mathbf{1}\}$ and $C(\widetilde{\mathbb{X}})$. First, as in Theorem 2.1 the projection of $\mathbf{y}$ onto span$\{\mathbf{1}\}$ is $\bar{y}\mathbf{1}$. Next, using the formula from Sect. 1.13, the projection of $\mathbf{y}$ onto $C(\widetilde{\mathbb{X}})$ is

$$\widetilde{\mathbb{X}}(\widetilde{\mathbb{X}}^T\widetilde{\mathbb{X}})^-\widetilde{\mathbb{X}}^T\mathbf{y} = \widetilde{\mathbb{X}}(\underbrace{\tfrac{1}{n}\widetilde{\mathbb{X}}^T\widetilde{\mathbb{X}}}_{\Sigma})^-\tfrac{1}{n}\widetilde{\mathbb{X}}^T\mathbf{y}.$$

The $j$th entry of $\frac{1}{n}\widetilde{\mathbb{X}}^T\mathbf{y}$ is

$$\begin{aligned}\tfrac{1}{n}\langle \mathbf{x}^{(j)} - \bar{x}^{(j)}\mathbf{1}, \mathbf{y}\rangle &= \tfrac{1}{n}\langle \mathbf{x}^{(j)} - \bar{x}^{(j)}\mathbf{1}, \mathbf{y} - \bar{y}\mathbf{1}\rangle \\ &= \sigma_{\mathbf{x}^{(j)},\mathbf{y}}.\end{aligned}$$

Finally, putting these together we obtain the orthogonal projection of $\mathbf{y}$ onto span$\{\mathbf{1}, \mathbf{x}^{(1)}, \ldots, \mathbf{x}^{(m)}\}$

$$\hat{\mathbf{y}} := \bar{y}\mathbf{1} + \widetilde{\mathbb{X}}\Sigma^- \begin{bmatrix} \sigma_{\mathbf{x}^{(1)},\mathbf{y}} \\ \vdots \\ \sigma_{\mathbf{x}^{(m)},\mathbf{y}} \end{bmatrix}.$$

Comparing this to the form of the prediction vector $a\mathbf{1} + \widetilde{\mathbb{X}}\mathbf{b}$, we can identify the optimal coefficients. ∎

By distributing the coefficients then regrouping the terms, we can derive the usual formulation of the least-squares hyperplane.

$$\begin{aligned} y &= \bar{y} + \hat{b}_1(x^{(1)} - \bar{x}^{(1)}) + \ldots + \hat{b}_m(x^{(m)} - \bar{x}^{(m)}) \\ &= (\bar{y} - \hat{b}_1\bar{x}^{(1)} - \ldots - \hat{b}_m\bar{x}^{(m)}) + \hat{b}_1 x^{(1)} + \ldots + \hat{b}_m x^{(m)} \end{aligned}$$

The least-squares hyperplane's intercept (generally called $\hat{b}_0$) is $\bar{y} - \hat{b}_1\bar{x}^{(1)} - \ldots - \hat{b}_m\bar{x}^{(m)}$, while the slopes (coefficients of the explanatory variables) are $\hat{b}_1, \ldots, \hat{b}_m$.

Compare the formulas for multiple linear regression to our earlier solution for simple linear regression (Theorem 2.2). If there is only one explanatory variable, the "inverse covariance matrix" is just the reciprocal of its variance; our earlier solution is indeed a special case of the multiple linear regression formula. And if there are no explanatory variables, the formula simplifies to *least-squares location regression* (Theorem 2.1). Of course, this makes the earlier theorems and proofs redundant, but they provided a valuable warm-up.

We will continue to draw variables pictures in $\mathbb{R}^n$, but now that we have more than three vectors of interest, we need to think carefully about our depictions. Let us look at the variables pictures in Fig. 2.3 and ask ourselves how accurate they are. We know that there exists *some* three-dimensional subspace that includes $\mathbf{y}$, $\hat{\mathbf{y}}$, and $\bar{y}\mathbf{1}$ (in addition to the origin). The question is, does span$\{\mathbf{x}^{(1)}, \ldots, \mathbf{x}^{(m)}\}$ intersect this subspace in a *line*? Assume that $\mathbf{y}$, $\hat{\mathbf{y}}$, and $\bar{y}\mathbf{1}$ are distinct vectors and that $\mathbf{1}$ is not in the span of the explanatory variables; typically, this is indeed the case. The vector $\hat{y} - \bar{y}\mathbf{1}$ is in the picture, and it is in the span of the explanatory vectors. Every scalar multiple of $\hat{y} - \bar{y}\mathbf{1}$ is also that span and in the three-dimensional subspace defining our picture, so there is *at least* a line of intersection. Could the intersection be *more* than a line? If it were a plane that did not include $\mathbf{1}$, then span$\{\mathbf{1}, \mathbf{x}^{(1)}, \ldots, \mathbf{x}^{(m)}\}$ would occupy all three dimensions in our picture and would therefore include $\mathbf{y}$, but that would contradict our assumption that $\hat{\mathbf{y}}$ and $\mathbf{y}$ are distinct.

### *2.1.4 Least-Squares Linear Regression*

Based on the intuition you are developing by working through this chapter, realize that *whenever the set of prediction functions under consideration comprises a subspace of* $\mathbb{R}^n$*, the prediction vector that minimizes the sum of squared residuals is precisely the orthogonal projection of the response vector onto that subspace.* The possible prediction functions comprise a subspace exactly when the form of the prediction function is linear in its free parameters. Furthermore, you are not limited to the explanatory variables you started with: you can use any functions of those variables to construct terms for your prediction functions. These observations lead us to a general formulation in this subsection.

**Exercise 2.6** Let $\mathbf{y} \in \mathbb{R}^n$ be a response variable and $\mathbf{x} \in \mathbb{R}^n$ be an explanatory variable. Consider fitting the response variable using quadratic functions of the explanatory variable:

$$\{f_{a,b,c}(x) = a + bx + cx^2 : a, b, c \in \mathbb{R}\}.$$

Show that the set of possible prediction vectors is a subspace of $\mathbb{R}^n$.

*Solution* Let $f_{a,b,c}(\mathbf{x})$ denote the vector of predictions $(f_{a,b,c}(x_1), \ldots, f_{a,b,c}(x_n))$. With $\mathbf{x}^2$ representing the vector of squared explanatory values, an arbitrary linear combination of two arbitrary vectors of predicted values is

$$\begin{aligned}\alpha_1 f_{a_1,b_1,c_1}(\mathbf{x}) + \alpha_2 f_{a_2,b_2,c_2}(\mathbf{x}) &= \alpha_1(a_1\mathbf{1} + b_1\mathbf{x} + c_1\mathbf{x}^2) + \alpha_2(a_2\mathbf{1} + b_2\mathbf{x} + c_2\mathbf{x}^2) \\ &= (\alpha_1 a_1 + \alpha_2 a_2)\mathbf{1} + (\alpha_1 b_1 + \alpha_2 b_2)\mathbf{x} + (\alpha_1 c_1 + \alpha_2 c_2)\mathbf{x}^2 \\ &= f_{\alpha_1 a_1 + \alpha_2 a_2, \alpha_1 b_1 + \alpha_2 b_2, \alpha_1 c_1 + \alpha_2 c_2}(\mathbf{x})\end{aligned}$$

which is another possible vector of predicted values that can be achieved using a quadratic function. In fact, we can see that the set of possible predictions is exactly the span of $\mathbf{1}$, $\mathbf{x}$, and $\mathbf{x}^2$. ♦

**Definition (Least-Squares Linear Fit)** Let $\mathbf{x}_1, \ldots, \mathbf{x}_n \in \mathbb{R}^m$ be $n$ observations of $m$ *explanatory* variables, and let $y_1, \ldots, y_n \in \mathbb{R}$ be corresponding values of a *response* variable. Given functions $g_1, \ldots, g_d$ that map from $\mathbb{R}^m$ to $\mathbb{R}$, the **least-squares linear fit** has the equation

$$y = \hat{c}_1 g_1(\mathbf{x}) + \ldots + \hat{c}_d g_d(\mathbf{x}),$$

where the coefficients $\hat{c}_1, \ldots, \hat{c}_d \in \mathbb{R}$ minimize the sum of squared residuals when each $y_i$ is predicted by a linear combination of functions of the explanatory variables,

$$\begin{aligned}(\hat{c}_1, \ldots, \hat{c}_d) := \underset{(c_1,\ldots,c_d)\in\mathbb{R}^d}{\operatorname{argmin}} \sum_i (y_i - [c_1 g_1(x_i^{(1)}, \ldots, x_i^{(m)}) + \ldots \\ + c_d g_d(x_i^{(1)}, \ldots, x_i^{(m)})])^2.\end{aligned}$$

The vector of residuals when using a prediction function of the above form is $\mathbf{y} - \mathbb{M}\mathbf{c}$ where

$$\mathbb{M} := \begin{bmatrix} g_1(\mathbf{x}_1) & \cdots & g_d(\mathbf{x}_1) \\ \vdots & \ddots & \vdots \\ g_1(\mathbf{x}_n) & \cdots & g_d(\mathbf{x}_n) \end{bmatrix}$$

is called the *design matrix* or *model matrix*. The sum of squared residuals equals $\|\mathbf{y} - \mathbb{M}\mathbf{c}\|^2$.

**Theorem 2.4 (Least-Squares Linear Regression)** *Let* $\mathbf{x}_1, \ldots, \mathbf{x}_n \in \mathbb{R}^m$, *and let* $y_1, \ldots, y_n \in \mathbb{R}$. *The objective function*

$$f(c_1, \ldots, c_d) = \sum_i (y_i - [c_1 g_1(\mathbf{x}_i) + \ldots + c_d g_d(\mathbf{x}_i)])^2$$

*is minimized by the entries of* $\hat{\mathbf{c}} := (\mathbb{M}^T \mathbb{M})^- \mathbb{M}^T \mathbf{y}$, *where* $\mathbb{M}$ *is the* design matrix *which has as its* $(i, j)$ *entry* $g_j(\mathbf{x}_i)$.

*Proof* The objective function can be expressed as $\|\mathbf{y} - \mathbb{M}\mathbf{c}\|^2$ which is minimized if $\mathbf{c}$ is $(\mathbb{M}^T \mathbb{M})^- \mathbb{M}^T \mathbf{y}$, as we discovered in Sect. 1.13. ■

The regression tasks covered in the previous sections are also cases of this more general least-squares linear regression. For example, including an intercept term means simply having a function that maps to 1 regardless of the values of its arguments, creating a **1** column in the design matrix.[13] Our earlier approaches to those derivations were more complicated, but the solutions we found are more interpretable and will aid our analyses in later chapters.

Exercises 2.6 and 2.7 reinforce the point that *linear* regression does not refer to the response variable being predicted by a linear function of the explanatory variable but rather by a linear function of the free parameters.

**Exercise 2.7** Let $\mathbf{y} \in \mathbb{R}^n$ be a response variable vector and $\mathbf{x} \in \mathbb{R}^n$ be an explanatory variable vector. Consider predicting the response variable by using quadratic functions of the explanatory variable:

$$\{f_{a,b,c}(x) = a + bx + cx^2 : a, b, c \in \mathbb{R}\}.$$

Explain how to find the coefficients $(\hat{a}, \hat{b}, \hat{c})$ of the quadratic function that minimizes the sum of squared residuals.

*Solution* With $\mathbf{x}^2$ representing the vector of squared explanatory values, we can use the design matrix

$$\mathbb{M} := \begin{bmatrix} | & | & | \\ \mathbf{1} & \mathbf{x} & \mathbf{x}^2 \\ | & | & | \end{bmatrix}.$$

[13] The least-squares coefficient of that column was identified as $\hat{b}_0$ in Sect. 2.1.3, while the least-squares coefficients of the $\mathbf{x}^{(1)}, \ldots, \mathbf{x}^{(m)}$ are exactly the same as the coefficients calculated for their centered versions.

According to Theorem 2.4, the least-squares coefficients are $(\hat{a}, \hat{b}, \hat{c}) = (\mathbb{M}^T\mathbb{M})^{-}\mathbb{M}^T\mathbf{y}$. ♦

Often we will be interested in the orthogonal projection of $\mathbf{y}$ onto $C(\mathbb{M})$ as well as the orthogonal projection onto a smaller subspace $\mathcal{S} \subseteq C(\mathbb{M})$ as depicted in Fig. 2.4. How accurate is this figure? We know that there exists *some* three-dimensional subspace that includes $\mathbf{y}$, $\hat{\mathbf{y}}$, and $\check{\mathbf{y}}$. But does $\mathcal{S}$ intersect this subspace in a *line*, and does $C(\mathbb{M})$ intersect it in a plane? Assume that $\mathbf{y}$, $\check{\mathbf{y}}$ and $\hat{\mathbf{y}}$ are distinct vectors. $\{a\check{\mathbf{y}} : a \in \mathbb{R}\} \subseteq \mathcal{S}$ is in our three-dimensional pictures, so the intersection with $\mathcal{S}$ is at least a line. And $\text{span}\{\check{\mathbf{y}}, \hat{\mathbf{y}}\} \subseteq C(\mathbb{M})$ is a plane that includes the line $\{a\check{\mathbf{y}} : a \in \mathbb{R}\}$. If the intersection with $C(\mathbb{M})$ was three-dimensional, then the entire picture would be in $C(\mathbb{M})$ which contradicts the assumption that $\hat{\mathbf{y}}$ and $\mathbf{y}$ are distinct. Because we have assumed $\check{\mathbf{y}} \neq \hat{\mathbf{y}}$, we can conclude that this plane is not in $\mathcal{S}$. Finally, if the intersection of this three-dimensional perspective with $\mathcal{S}$ had another dimension (outside of $\text{span}\{\check{\mathbf{y}}, \hat{\mathbf{y}}\}$), this would also imply that the intersection with $C(\mathbb{M})$ is three-dimensional.

What if the vectors and subspaces of interest are not linearly independent in the ways we assumed above? In that case, the true picture *may* differ from our depiction. For example, if $\mathbf{y} \in C(\mathbb{M})$, then $\hat{\mathbf{y}} = \mathbf{y}$. It is good to be aware of these types of possibilities, but our drawings represent the typical case of linearly independence. And even when linear dependence causes the picture to be imperfect, the conclusions drawn from the picture are usually still valid. To be completely thorough, one must think through each possible way in which the picture can differ from reality and think about whether intuition derived from the picture holds or fails in those cases.

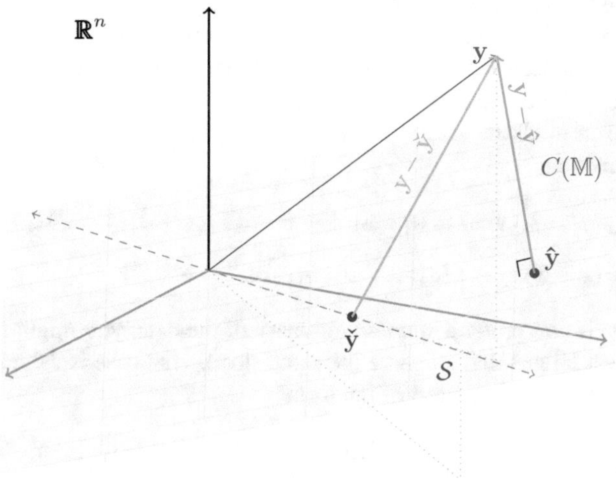


**Fig. 2.4** A response variable vector $\mathbf{y}$ in $\mathbb{R}^n$, its orthogonal projection $\check{\mathbf{y}}$ onto a subspace $\mathcal{S} \subseteq C(\mathbb{M})$, and its orthogonal projection $\hat{\mathbf{y}}$ onto $C(\mathbb{M})$ inhabit a three-dimensional subspace of $\mathbb{R}^n$. In general, $\mathcal{S}$ intersects this three-dimensional subspace in a line, and $C(\mathbb{M})$ intersects it in a plane

**Exercise 2.8** Let $\hat{\mathbf{y}}$ be the orthogonal projection of $\mathbf{y}$ onto $C(\mathbb{M})$. Explain why $(\mathbb{M}^T\mathbb{M})^-\mathbb{M}^T\hat{\mathbf{y}}$ must be equal to $(\mathbb{M}^T\mathbb{M})^-\mathbb{M}^T\mathbf{y}$.

*Solution* There is an intuitive explanation for this. You can think of $(\mathbb{M}^T\mathbb{M})^-\mathbb{M}^T$ as the matrix that maps any vector in $\mathbb{R}^n$ to the (minimum norm) coefficients of the columns of $\mathbb{M}$ that lead to the orthogonal projection of that vector onto $C(\mathbb{M})$. Because the orthogonal projection of $\hat{\mathbf{y}}$ onto $C(\mathbb{M})$ is exactly the same as the orthogonal projection of $\mathbf{y}$ onto $C(\mathbb{M})$ (namely, both are $\hat{\mathbf{y}}$), the coefficients leading to this orthogonal projection must be the same. ♦

#### 2.1.4.1 Regression with Categorical Explanatory Variables

Finally, let us see how least-squares linear regression can be applied with categorical explanatory variables. With a single categorical explanatory variable $\mathbf{z}$ taking the values $\{1, \ldots, k\}$, a natural set of functions for predicting the data is[14]

$$\{f_{a_1,\ldots,a_k}(z) = a_1\mathbb{1}_{z=1} + \ldots + a_k\mathbb{1}_{z=k} : (a_1, \ldots, a_k) \in \mathbb{R}^k\}.$$

In other words, all the observations of group $j$ will be predicted by some number $b_j$. We know from Sect. 2.1.1 that the least-squares coefficients must be simply the corresponding groups' averages. What does the corresponding design matrix look like? With $z_i = j$, the $i$th row of the design matrix has a 1 in column $j$ and zeros elsewhere. Notice that the columns are all orthogonal to each other, so linear independence of the columns always holds in this context. It is also important to realize that the sum of the columns is $\mathbf{1}$, so the upcoming Theorem 2.5 will be valid.

Linear regression can use any number of categorical and any number of quantitative explanatory variables. The form of the prediction function depends on how you want to allow the variables to *interact*. For example, if $\mathbf{x}$ is a quantitative explanatory variable, $\mathbf{z}$ is a categorical explanatory variable, and $\mathbf{y}$ is a quantitative response variable, you could calculate a separate least-squares line for each group. Written in terms of all the variables together, this is called an *interactions* fit and has prediction functions

$$\{f_{a_1,\ldots,a_k,b_1,\ldots,b_k}(x, z) = a_1\mathbb{1}_{z=1} + \cdots + a_k\mathbb{1}_{z=k} + b_1(x - \bar{x})\mathbb{1}_{z=1} + \cdots$$
$$+ b_k(x - \bar{x})\mathbb{1}_{z=k} : (a_1, \ldots, a_k, b_1, \ldots, b_k) \in \mathbb{R}^{2k}\}.$$

Alternatively, if you prefer a *simpler* summary of the data, you might instead want to find the best $k$ lines that all share the same slope. That case is called an *additive* fit and uses prediction functions of the form

$$\{f_{a_1,\ldots,a_k,b}(x, z) = a_1\mathbb{1}_{z=1} + \ldots + a_k\mathbb{1}_{z=k} + b(x - \bar{x}) : (a_1, \ldots, a_k, b) \in \mathbb{R}^{k+1}\}.$$

[14]The notation $\mathbb{1}_{\text{condition}}$ is called an indicator function; it takes the value 1 if its condition is true and 0 otherwise.

## 2.2 Sums of Squares

With the variables picture in mind, we derived least-squares coefficients and predictions by understanding them in terms of orthogonal projection theory which was covered at length in Chap. 1. Closely related to this, another benefit of the variables picture is that it enables us to easily derive a variety of useful decompositions of certain sums of squares by observing right triangles in the picture and simply invoking the Pythagorean identity.

### *2.2.1 ANOVA Decomposition*

**Theorem 2.5 (ANOVA Decomposition)** *Let* $\hat{\mathbf{y}} = (\hat{y}_1, \ldots, \hat{y}_n)$ *be the orthogonal projection of* $\mathbf{y} = (y_1, \ldots, y_n)$ *onto* $\mathcal{S}$. *If* $\mathbf{1} \in \mathcal{S}$, *then*

$$\sum_i (y_i - \bar{y})^2 = \sum_i (\hat{y}_i - \bar{y})^2 + \sum_i (y_i - \hat{y}_i)^2.$$

*Proof* Because $\hat{\mathbf{y}}$ and $\bar{y}\mathbf{1}$ are both in $\mathcal{S}$, so is their difference $\hat{\mathbf{y}} - \bar{y}\mathbf{1}$. We can express $\mathbf{y} - \bar{y}\mathbf{1}$ as the sum of the vectors $\hat{\mathbf{y}} - \bar{y}\mathbf{1}$ and $\mathbf{y} - \hat{\mathbf{y}}$. The first vector is in $\mathcal{S}$, whereas the second is orthogonal to $\mathcal{S}$, so the two vectors are orthogonal to each other (see Fig. 2.5). Rewriting the sums as squared norms of vectors, then invoking the Pythagorean identity,

$$\begin{aligned}\sum_i (y_i - \bar{y})^2 &= \|\mathbf{y} - \bar{y}\mathbf{1}\|^2 \\ &= \|\hat{\mathbf{y}} - \bar{y}\mathbf{1}\|^2 + \|\mathbf{y} - \hat{\mathbf{y}}\|^2 \\ &= \sum_i (\hat{y}_i - \bar{y})^2 + \sum_i (y_i - \hat{y}_i)^2.\end{aligned}$$

■

The terms of the ANOVA decomposition are sometimes called the *total sum of squares*, *regression sum of squares*, and *residual sum of squares*, respectively. The acronym *ANOVA* stands for *an*alysis *o*f *va*riance. By dividing both sides of the equation by $n$, you can interpret the quantities as the empirical variance of the *response* vector $\mathbf{y}$, the empirical variance of *regression* vector $\hat{\mathbf{y}}$ (which has mean $\bar{y}$ according to Exercise 1.50), and the empirical variance of the (least-squares) *residual* vector $\mathbf{y} - \hat{\mathbf{y}}$ (which has mean zero because it is orthogonal to $\mathbf{1}$).

$$\sigma_{\mathbf{y}}^2 = \sigma_{\hat{\mathbf{y}}}^2 + \sigma_{\mathbf{y}-\hat{\mathbf{y}}}^2$$

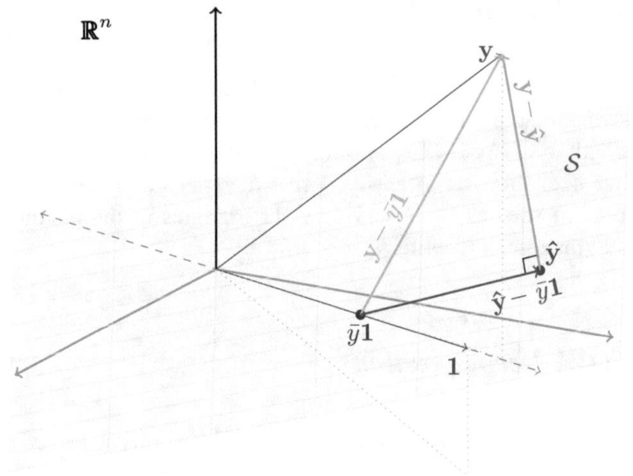


**Fig. 2.5** The ANOVA decomposition is the Pythagorean identity applied to the right triangle pictured

The fraction $\frac{\sigma^2_{\hat{\mathbf{y}}}}{\sigma^2_{\mathbf{y}}}$ is called the $R^2$ of the regression; it represents the proportion of the total sum of squares that was "explained" by the explanatory variables. Notice that this fraction (also written $\frac{\|\hat{\mathbf{y}}-\bar{y}\mathbf{1}\|^2}{\|\mathbf{y}-\bar{y}\mathbf{1}\|^2}$) is the squared cosine of the angle between $\mathbf{y}-\bar{y}\mathbf{1}$ and $\hat{\mathbf{y}}-\bar{y}\mathbf{1}$. When there is only one explanatory variable, $\hat{\mathbf{y}}-\bar{y}\mathbf{1}$ is in the direction of $\mathbf{x}-\bar{x}\mathbf{1}$, as we can see in the least-squares line equation. So in that case $R^2$ is the squared cosine of the angle between $\mathbf{y}-\bar{y}\mathbf{1}$ and $\mathbf{x}-\bar{x}\mathbf{1}$ which, referring back to the definition of correlation in Sect. 2.1.2, means that $R^2$ is the squared correlation[15] between $\mathbf{y}$ and $\mathbf{x}$.

In the case of one categorical explanatory variable, the ANOVA decomposition (Theorem 2.5) is often written by summing over the groups rather than the observations. With $k$ groups, we let $\bar{y}_1, \ldots, \bar{y}_k$ represent the groups' means and $n_1, \ldots, n_k$ represent the number of observations in each group. The least-squares procedure predicts every observation by its group's mean, so

$$\begin{aligned}\sum_i (y_i - \bar{y})^2 &= \sum_i (\hat{y}_i - \bar{y})^2 + \sum_i (y_i - \hat{y}_i)^2 \\ &= \sum_i (\bar{y}_{x_i} - \bar{y})^2 + \sum_i (y_i - \bar{y}_{x_i})^2\end{aligned}$$

[15]Galton used the letter $R$ for the "regression coefficient" in the least-squares equation for the standardized variables, which we now call the *correlation*. In simple linear regression, $R^2$ is exactly the squared correlation, which is why the statistic is called $R^2$. Note that with more than one explanatory variable, this interpretation no longer works.

$$= \sum_{j=1}^{k} n_j(\bar{y}_j - \bar{y})^2 + \sum_{j=1}^{k} \sum_{i:x_i=j} (y_i - \bar{y}_j)^2. \quad (2.1)$$

When there are multiple variables, the regression sum of squares can be further decomposed. Note that the logic behind Exercise 2.9 can be extended to any number of nested subspaces.

**Exercise 2.9** Suppose $\hat{\mathbf{y}}$ is the orthogonal projection of $\mathbf{y}$ onto $\mathcal{S}$, $\check{\mathbf{y}}$ is the orthogonal projection of $\mathbf{y}$ onto $\mathcal{S}_0 \subseteq \mathcal{S}$, and that $\mathbf{1} \in \mathcal{S}_0$. Explain why

$$\|\hat{\mathbf{y}} - \bar{y}\mathbf{1}\|^2 = \|\check{\mathbf{y}} - \bar{y}\mathbf{1}\|^2 + \|\hat{\mathbf{y}} - \check{\mathbf{y}}\|^2.$$

*Solution* The vector $\check{\mathbf{y}}$ is defined to be the orthogonal projection of $\mathbf{y}$ onto $\mathcal{S}_0$. However, it is also the orthogonal projection of $\hat{\mathbf{y}}$ onto $\mathcal{S}_0$ because according to Exercise 1.50, orthogonal projection onto $\mathcal{S}$ followed by orthogonal projection onto $\mathcal{S}_0$ lands you at the exact same vector that a single orthogonal projection onto $\mathcal{S}_0$ does. Likewise, $\bar{y}\mathbf{1}$ is the orthogonal projection of $\check{\mathbf{y}}$ onto $\mathbf{1}$. Invoke the ANOVA decomposition with $\hat{\mathbf{y}}$ playing the role of the response variable. ♦

### 2.2.2 *Contrasts*

We will conclude this chapter by working through another important decomposition of the regression sum of squares $\|\hat{\mathbf{y}} - \bar{y}\mathbf{1}\|^2$ in the context of categorical explanatory variables. The results from this discussion will be used in Chap. 6.

Let $\mathbf{x}$ be a categorical explanatory variable taking values $\{1, \ldots, k\}$. A corresponding design matrix can be defined by

$$\mathbb{M} = \begin{bmatrix} \mathbb{1}_{x_1=1} & \cdots & \mathbb{1}_{x_1=k} \\ \vdots & \ddots & \vdots \\ \mathbb{1}_{x_n=1} & \cdots & \mathbb{1}_{x_n=k} \end{bmatrix}.$$

Suppose every group having $n/k$ observations. (When all groups have the same number of observations, they are called *balanced*.) In this context, any vector $\mathbf{w} = (w_1, \ldots, w_k) \in \mathbb{R}^k$ that has mean zero is called a *contrast* with its $k$ entries corresponding to the $k$ groups. Consider the vector $\mathbb{M}\mathbf{w} \in \mathbb{R}^n$. Its $i$th entry is equal to the contrast value for the group that $x_i$ belongs to. We can also express that observation as $\mathbb{M}\mathbf{w} = (w_{x_1}, \ldots, w_{x_n})$.

Let $\mathbf{w}^{(1)} = (w_1^{(1)}, \ldots, w_k^{(1)})$ and $\mathbf{w}^{(2)} = (w_1^{(2)}, \ldots, w_k^{(2)})$ in $\mathbb{R}^k$. The inner product of the corresponding vectors $\mathbb{M}\mathbf{w}^{(1)}$ and $\mathbb{M}\mathbf{w}^{(2)}$ in $\mathbb{R}^n$ is proportional to the inner product of $\mathbf{w}^{(1)}$ and $\mathbf{w}^{(2)}$.

$$\begin{aligned}\langle \mathbb{M}\mathbf{w}^{(1)}, \mathbb{M}\mathbf{w}^{(2)}\rangle &= \sum_{i=1}^{n} w_{x_i}^{(1)} w_{x_i}^{(2)} \\ &= \sum_{j=1}^{k} (n/k) w_j^{(1)} w_j^{(2)} \\ &= \tfrac{n}{k}\langle \mathbf{w}^{(1)}, \mathbf{w}^{(2)}\rangle \end{aligned}$$

because each group appears exactly $n/k$ times in the balanced dataset. From this, we can see that if $\mathbf{w}^{(1)}$ and $\mathbf{w}^{(2)}$ are orthogonal to each other, then so are $\mathbb{M}\mathbf{w}^{(1)}$ and $\mathbb{M}\mathbf{w}^{(2)}$. Secondly, if $\mathbf{w}^{(l)}$ has mean zero (is orthogonal to $\mathbf{1}$), then $\mathbb{M}\mathbf{w}^{(l)}$ also has mean zero because it is orthogonal to $\mathbb{M}\mathbf{1} = \mathbf{1} \in \mathbb{R}^n$. Third, the squared norms of corresponding vectors have the simple relationship $\|\mathbb{M}\mathbf{w}^{(l)}\|^2 = \frac{n}{k}\|\mathbf{w}^{(l)}\|^2$.

Suppose $\mathbf{w}^{(1)}, \ldots, \mathbf{w}^{(k-1)} \in \mathbb{R}^k$ are contrasts that are all orthogonal to each other. Then the set $\{\mathbf{1}, \mathbf{w}^{(1)}, \ldots, \mathbf{w}^{(k-1)}\}$ is an orthogonal basis for $\mathbb{R}^k$, and their corresponding vectors $\{\mathbf{1}, \mathbb{M}\mathbf{w}^{(1)}, \ldots, \mathbb{M}\mathbf{w}^{(k-1)}\}$ are an orthogonal basis for the $k$-dimensional subspace $C(\mathbb{M}) \subseteq \mathbb{R}^n$.

Based on Exercise 1.44, the regression vector $\hat{\mathbf{y}} - \bar{y}\mathbf{1}$ is the orthogonal projection of $\mathbf{y}$ onto the orthogonal complement of $\mathbf{1}$ in $C(\mathbb{M})$. This same orthogonal projection can be readily expressed using the orthonormal basis $\{\frac{\mathbb{M}\mathbf{w}^{(1)}}{\|\mathbb{M}\mathbf{w}^{(1)}\|}, \ldots, \frac{\mathbb{M}\mathbf{w}^{(k-1)}}{\|\mathbb{M}\mathbf{w}^{(k-1)}\|}\}$ for that subspace.

$$\hat{\mathbf{y}} - \bar{y}\mathbf{1} = \left\langle \mathbf{y}, \frac{\mathbb{M}\mathbf{w}^{(1)}}{\|\mathbb{M}\mathbf{w}^{(1)}\|}\right\rangle \frac{\mathbb{M}\mathbf{w}^{(1)}}{\|\mathbb{M}\mathbf{w}^{(1)}\|} + \ldots + \left\langle \mathbf{y}, \frac{\mathbb{M}\mathbf{w}^{(k-1)}}{\|\mathbb{M}\mathbf{w}^{(k-1)}\|}\right\rangle \frac{\mathbb{M}\mathbf{w}^{(k-1)}}{\|\mathbb{M}\mathbf{w}^{(k-1)}\|}$$

Its squared length is

$$\|\hat{\mathbf{y}} - \bar{y}\mathbf{1}\|^2 = \left\langle \mathbf{y}, \frac{\mathbb{M}\mathbf{w}^{(1)}}{\|\mathbb{M}\mathbf{w}^{(1)}\|}\right\rangle^2 + \ldots + \left\langle \mathbf{y}, \frac{\mathbb{M}\mathbf{w}^{(k-1)}}{\|\mathbb{M}\mathbf{w}^{(k-1)}\|}\right\rangle^2.$$

Let us work out an alternative expression for the inner product between $\mathbf{y}$ and $\mathbb{M}\mathbf{w}^{(l)}$ by summing over the groups:

$$\begin{aligned}\langle \mathbf{y}, \mathbb{M}\mathbf{w}^{(l)}\rangle &= \sum_{i} w_{x_i}^{(l)} y_i \\ &= \sum_{j=1}^{k} w_j^{(l)} \underbrace{\sum_{i:x_i=j} y_i}_{(n/k)\bar{y}_j} \\ &= (n/k)\sum_{j=1}^{k} w_j^{(l)} \bar{y}_j. \end{aligned}$$

Thus we can derive a simpler expression for the square of the coefficient of $\mathbf{y}$ with respect to $\frac{\mathbb{M}\mathbf{w}^{(l)}}{\|\mathbb{M}\mathbf{w}^{(l)}\|}$:

$$\left\langle \mathbf{y}, \frac{\mathbb{M}\mathbf{w}^{(l)}}{\|\mathbb{M}\mathbf{w}^{(l)}\|} \right\rangle^2 = \frac{\langle \mathbf{y}, \mathbb{M}\mathbf{w}^{(l)} \rangle^2}{\|\mathbb{M}\mathbf{w}^{(l)}\|^2}$$

$$= \frac{(n/k)^2 (\sum_j w_j^{(l)} \bar{y}_j)^2}{(n/k)\|\mathbf{w}^{(l)}\|^2}$$

$$= \frac{(n/k)(\sum_j w_j^{(l)} \bar{y}_j)^2}{\sum_j (w_j^{(l)})^2}. \tag{2.2}$$

Section 6.2.4.1 will make use of this for hypothesis testing.

In Chap. 4, we will make certain *modeling assumptions* (statements about a probability distribution generating the data), and see how to incorporate them into our visualizations. But first, Chap. 3 will discuss random vectors and some of their key properties.

# Chapter 3
# Background: Random Vectors

So far, we have only worried about approximating or fitting data; we have not asked *why* the data looks like it does. Moving forward, we'll use probabilistic modeling to consider possible mechanisms generating the data. In this chapter, we'll briefly learn how to work with random vectors, the building blocks of probabilistic modeling.

As with Chap. 1, this chapter builds from the ground up and is largely self-contained. Section 3.2 is fairly abstract but as usual the most detailed work is hidden away in the solutions to the green exercises.

We are going to take a rather different approach to probability and expectation than most books. Most notably, we are not going to care about whether functions on a sample space are "measurable" or not. In my view, expectation makes sense without reference to measurability; measurability primarily helps answer the follow-up question of whether or not the expectation is uniquely determined. However, the identities and inequalities that we derive involving expectations can be asserted regardless of whether each expectation has a uniquely determined value or not. Appendix C of Brinda (2018) explains this interpretation.

## 3.1 Probability

**Definition (Sample Space)** A **sample space** $\Omega$ is a set that has *probabilities* assigned to a selection of its subsets which are called *events*. The empty set has probability 0, and $\Omega$ has probability 1. If $E$ has probability $p$, its *complement* has probability $1 - p$. If $E_1, E_2, \ldots$ are *disjoint* subsets with probabilities $p_1, p_2, \ldots$, their union has probability $\sum_i p_i$.

The probabilities of certain additional subsets that are not designated as events might be *implied*. If $E_1 \subseteq E_2$ are events that have the same probability, and $S$ is a subset nested between them ($E_1 \subseteq S \subseteq E_2$), then $S$ should also be considered to have that probability.



W. D. Brinda, *Visualizing Linear Models*,
https://doi.org/10.1007/978-3-030-64167-2_3

The probability assigned to $E$ is denoted $\mathbb{P}\, E$.

**Exercise 3.1** Explain why the sum of the probabilities of $E_1$ and $E_2$ is no greater than the probability of their union.

**Definition (Random Element)** A **random element** is a function whose domain is a *sample space*.

Crucially, *a random element maps probabilities from its domain to its range*. Let the notation $f(S)$ denote the set $\{f(x) : x \in S\}$ which is called the *image* of $S$ (under $f$).

**Definition (Distribution)** The **distribution** of a random element $X$ assigns to $X(S)$ a probability equal to the probability of $S$, for every subset $S$ of the sample space that is assigned a probability.

It is more common to start with a subset $S$ in the random element's range and ask what probability it is assigned by the distribution. Such a probability $\mathbb{P}\{\omega : X(\omega) \in S\}$ will generally be written as $\mathbb{P}\{X \in S\}$.

**Definition (Random Vector)** A random element that maps to a vector space with an inner product will be called a **random vector**. A random element that maps to $\mathbb{R}$ will be called a **random variable**.

Note that a *random variable* is an $\mathbb{R}$*-valued random vector*.

The usual notation for random variables often leads students to forget that they are functions. For instance, with random variables $X$ and $Y$ defined on a sample space $\Omega$ and $a, b \in \mathbb{R}$, consider the random variable $Z := aX + Y + b$. It is a function from $\Omega$ to $\mathbb{R}$. One can make that more clear by writing it with a dummy variable, $Z(\omega) = aX(\omega) + Y(\omega) + b$ for all $\omega \in \Omega$. However, often the fact that they are functions is not important if you know their distributions; many questions can be answered simply by making reference to the distributions.

Results from Chaps. 1 and 2 that were true for arbitrary vectors also hold *pointwise* for random vectors, meaning that they are true at every point in the sample space. For example, let $\mathbf{Y}$ be a random vector, let $\widehat{\mathbf{Y}}$ be its orthogonal projection onto a subspace, and let $\mathbf{v}$ be a non-random vector in that subspace. Keep in mind that $\mathbf{Y}$ and $\widehat{\mathbf{Y}}$ are both function on the sample space $\Omega$ and for any $\omega \in \Omega$, $\widehat{\mathbf{Y}}(\omega)$ is determined by $\mathbf{Y}(\omega)$. *For each* $\omega \in \Omega$, the Pythagorean identity implies that

$$\|\mathbf{Y}(\omega) - \mathbf{v}\|^2 = \|\mathbf{Y}(\omega) - \widehat{\mathbf{Y}}(\omega)\|^2 + \|\widehat{\mathbf{Y}}(\omega) - \mathbf{v}\|^2.$$

That is what we mean if we write as short-hand

$$\|\mathbf{Y} - \mathbf{v}\|^2 = \|\mathbf{Y} - \widehat{\mathbf{Y}}\|^2 + \|\widehat{\mathbf{Y}} - \mathbf{v}\|^2;$$

*the two sides of the equation are the same function.*

**Definition (Discrete Random Element)** A random element $X$ is called **discrete** if its distribution has only countably many possible values $x_1, x_2, \ldots$, that is, $\sum_i p(x_i) = 1$ where $p(x_i) := \mathbb{P}\{X = x_i\}$. The function $p$ is called the **probability mass function** (pmf) of the distribution of $X$, or simply the pmf of $X$.

**Definition (Continuous Random Vector)** We will say that an $\mathbb{R}^d$-valued random vector $\mathbf{X}$ is **continuous** if there exists a function $p$ such that

$$\mathbb{P}\{\mathbf{X} \in E\} = \int_E p(\mathbf{x})d\mathbf{x}$$

for every event $E \subseteq \mathbb{R}^d$. The function $p$ is called a **probability density function** (pdf) of the distribution of $\mathbf{X}$, or simply a pdf of $\mathbf{X}$.

**Definition (Cumulative Distribution Function)** Given a distribution on $\mathbb{R}$, its **cumulative distribution function** evaluated at $t \in \mathbb{R}$ equals the probability that it assigns to $(-\infty, t]$.

## 3.2 Expectation

Let $\mathbb{1}_A$ to denote the *indicator function* for the set $A$; $\mathbb{1}_A(\omega)$ equals 1 if $\omega \in A$ and it equals 0 otherwise. As short-hand, we'll sometimes write a condition as the subscript rather than a set; for instance, $\mathbb{1}_{\mathbf{X} \in S}$ is short-hand for $\mathbb{1}_{\{\omega : \mathbf{X}(\omega) \in S\}}$.

**Definition (Expectation)** Let $\mathbf{X}$ and $\mathbf{Y}$ be random vectors, and let $Z$ be a random variable. The **expectation** operator (denoted $\mathbb{E}$) satisfies the following properties:

(i) $\mathbb{E}\mathbb{1}_{\mathbf{X} \in S}$ equals the probability of $S$ assigned by the distribution of $\mathbf{X}$
(ii) $\mathbb{E}\langle \mathbf{X}, \mathbf{v}\rangle = \langle \mathbb{E}\mathbf{X}, \mathbf{v}\rangle$ for all $\mathbf{v}$ if $\mathbb{E}\mathbf{X}$ is well-defined and finite
(iii) $\mathbb{E}(\mathbf{X} + \mathbf{Y}) = \mathbb{E}\mathbf{X} + \mathbb{E}\mathbf{Y}$ if $\mathbb{E}\mathbf{X}$ and $\mathbb{E}\mathbf{Y}$ are well-defined
(iv) $\mathbb{E}|Z| = \sup\{\mathbb{E}Y : Y \leq |Z|\}$
(v) $\mathbb{E}Z = \mathbb{E}[Z\mathbb{1}_{Z>0}] - \mathbb{E}[-Z\mathbb{1}_{Z<0}]$ (this difference is undefined if the two expectations are infinite with the same sign).

Notice the self-referential subtleties of the above definition. The first property defines expectation for indicator functions of events. The next three properties enable its extension to non-negative random variables (Pollard 2002 Sec 2.4). The final property extends it to all random variables. Next, property (ii) *uniquely* defines an expectation[1] for random vectors mapping to a vector space with an inner product. Exercise 3.3 shows that this expectation indeed satisfies property (iii) as well.

If a vector $\mathbb{E}\mathbf{X}$ exists, it is unique based on Exercise 1.30 and can be identified by property (ii). In other words, if $\mathbf{x}$ satisfies

[1]The expectation of a random vector as we define it here is also called the *Pettis expectation* in the literature.

$$\mathbb{E}\langle \mathbf{X}, \mathbf{v}\rangle = \langle \mathbf{x}, \mathbf{v}\rangle$$

for all $\mathbf{v}$, then $\mathbf{x}$ must be the expectation $\mathbb{E}\mathbf{X}$.

**Exercise 3.2** Let $\mathbf{X}$ be a random vector mapping to the complex plane with the representation $\mathbf{X} = Y + iZ$ where $Y$ and $Z$ are random variables. Verify that $\mathbb{E}Y + i\mathbb{E}Z$ is the expectation of $\mathbf{X}$ by checking property (ii), assuming that property (iii) holds for *random variables*.

**Exercise 3.3** Suppose $\mathbf{x}$ satisfies $\mathbb{E}\langle \mathbf{X}, \mathbf{v}\rangle = \langle \mathbf{x}, \mathbf{v}\rangle$ and $\mathbf{y}$ satisfies $\mathbb{E}\langle \mathbf{Y}, \mathbf{v}\rangle = \langle \mathbf{y}, \mathbf{v}\rangle$ for all $\mathbf{v}$. In order to justify an implicit claim in our definition of expectation for random vectors, verify that

$$\mathbb{E}\langle \mathbf{X} + \mathbf{Y}, \mathbf{v}\rangle = \langle \mathbf{x} + \mathbf{y}, \mathbf{v}\rangle$$

for all $\mathbf{v}$. In other words, verify that the expectation of a sum is indeed the sum of the expectations when all expectations are defined by property (ii), assuming that property (iii) holds for *random variables*.

We have provided an extraordinarily *concise* definition of expectation that works both for random variables and for random vectors more generally—hopefully it is not so concise that it's inscrutable. Try not to get bogged down in the details. In practice, we'll be able to manipulate expectations easily by making use of properties (ii) and (iii) and the results of some of this section's exercises.

**Exercise 3.4** For a random vector $\mathbf{X}$ and scalar $a$, show that $\mathbb{E}a\mathbf{X} = a\mathbb{E}\mathbf{X}$.

**Exercise 3.5** Suppose the random vector $\mathbf{X}$ maps every point in the sample space to $\mathbf{w}$. Show that $\mathbb{E}\mathbf{X} = \mathbf{w}$.

**Exercise 3.6** Let $\mathbf{X}$ be a random vector and $\mathbf{v}$ be a non-random vector. Explain why $\mathbb{E}(\mathbf{X} + \mathbf{v}) = \mathbb{E}\mathbf{X} + \mathbf{v}$.

*Solution* The random vector $\mathbf{X} + \mathbf{v}$ maps any $\omega$ to $\mathbf{X}(\omega) + \mathbf{v}$; we are justified in treating $\mathbf{v}$ as if it is the random vector that maps every element of the sample space to the vector $\mathbf{v}$. By property (iii), $\mathbb{E}(\mathbf{X} + \mathbf{v}) = \mathbb{E}\mathbf{X} + \mathbb{E}\mathbf{v}$, and by Exercise 3.5, $\mathbb{E}\mathbf{v} = \mathbf{v}$. ♦

**Exercise 3.7** Suppose $\mathbb{E}\mathbf{X} = \mathbf{0}$. Show that the coordinate of $\mathbf{X}$ with respect to $\mathbf{u}$ has expectation 0.

*Solution*

$$\begin{aligned}\mathbb{E}\langle \mathbf{X}, \mathbf{u}\rangle &= \langle \underbrace{\mathbb{E}\mathbf{X}}_{\mathbf{0}}, \mathbf{u}\rangle \\ &= 0\end{aligned}$$

♦

Suppose $\mathbf{X}$ is a $\mathcal{V}$-valued random vector and $\mathbf{u}_1, \ldots, \mathbf{u}_m$ comprise an orthonormal basis for $\mathcal{V}$. The coordinate of $\mathbb{E}\mathbf{X}$ with respect to $\mathbf{u}_j$ is exactly the expectation of the coordinate of $\mathbf{X}$ with respect to $\mathbf{u}_j$, by the defining property of random vectors: $\langle \mathbb{E}\mathbf{X}, \mathbf{u}_j\rangle = \mathbb{E}\langle \mathbf{X}, \mathbf{u}_j\rangle$. For instance, suppose $\mathbf{X}$ is an $\mathcal{F}^n$-valued random vector. Using the standard basis to define coordinates, the expectation of $\mathbf{X} = (X_1, \ldots, X_n)$ is simply $\mathbb{E}\mathbf{X} = (\mathbb{E}X_1, \ldots, \mathbb{E}X_n)$; the coordinates of the expectation vector are the expectations of the coordinate random variables.

**Exercise 3.8** Let $\mathbf{X}$ be a random vector that maps to a real vector space with an inner product. Show that the expected squared length of $\mathbf{X}$ equals sum of the expected squares of its coordinates with respect to any orthonormal basis $\mathbf{u}_1, \ldots, \mathbf{u}_m$.

*Solution* This is a simple consequence of Parseval's identity.

$$\begin{aligned}\mathbb{E}\|\mathbf{X}\|^2 &= \mathbb{E}[\langle \mathbf{X}, \mathbf{u}_1\rangle^2 + \ldots + \langle \mathbf{X}, \mathbf{u}_m\rangle^2] \\ &= \mathbb{E}\langle \mathbf{X}, \mathbf{u}_1\rangle^2 + \ldots + \mathbb{E}\langle \mathbf{X}, \mathbf{u}_m\rangle^2\end{aligned}$$

♦

**Exercise 3.9** Let $\mathbf{X} = (X_1, \ldots, X_n)$ be a random vector, $\mathbf{v} = (v_1, \ldots, v_n)$ be a non-random vector, and $\mathbb{M}$ be an $n \times m$ matrix. Show that

$$\mathbb{E}(\mathbf{v} + \mathbb{M}\mathbf{X}) = \mathbf{v} + \mathbb{M}\mathbb{E}\mathbf{X}.$$

*Solution* From Exercise 3.6, $\mathbb{E}(\mathbf{v} + \mathbb{M}\mathbf{X}) = \mathbf{v} + \mathbb{E}\mathbb{M}\mathbf{X}$. Let $\mathbf{m}_1, \ldots, \mathbf{m}_n$ be the rows of $\mathbb{M}$. Putting the expectation into each coordinate of the vector,

$$\begin{aligned}\mathbb{E}\mathbb{M}\mathbf{X} &= \mathbb{E}\begin{bmatrix}\mathbf{m}_1^T\mathbf{X} \\ \vdots \\ \mathbf{m}_n^T\mathbf{X}\end{bmatrix} \\ &= \begin{bmatrix}\mathbb{E}\mathbf{m}_1^T\mathbf{X} \\ \vdots \\ \mathbb{E}\mathbf{m}_n^T\mathbf{X}\end{bmatrix} \\ &= \begin{bmatrix}\mathbf{m}_1^T\mathbb{E}\mathbf{X} \\ \vdots \\ \mathbf{m}_n^T\mathbb{E}\mathbf{X}\end{bmatrix} \\ &= \mathbb{M}\mathbb{E}\mathbf{X}.\end{aligned}$$

♦

**Exercise 3.10** Suppose $\mathbf{X}$ is a discrete random vector with probability mass function $p$ on $\{\mathbf{x}_1, \ldots, \mathbf{x}_n\}$. Show that $\mathbb{E}\mathbf{X} = \sum_i \mathbf{x}_i p(\mathbf{x}_i)$.

A generalization of Exercise 3.10 follows from essentially the same line of reasoning. If $X$ is a discrete random element with possible values $\{x_1, \ldots, x_n\}$ and $f$ is a vector-valued function, then

$$\mathbb{E}f(X) = \sum_i f(x_i)p(x_i).$$

This statement also holds with a *countable infinity* of possible values as long as $f$ is not too *irregular*; for example, if $f$ is a non-negative real-valued function, then the result follows from property (iv).

*Solution* The random vector can be represented by the sum

$$\mathbf{X}(\omega) = \mathbf{x}_1 \mathbb{1}_{\mathbf{X}(\omega)=\mathbf{x}_1} + \ldots + \mathbf{x}_n \mathbb{1}_{\mathbf{X}(\omega)=\mathbf{x}_n}.$$

Taking the expectation,

$$\begin{aligned}\mathbb{E}\mathbf{X} &= \mathbb{E}[\mathbf{x}_1 \mathbb{1}_{\mathbf{X}=\mathbf{x}_1} + \ldots + \mathbf{x}_n \mathbb{1}_{\mathbf{X}=\mathbf{x}_n}] \\ &= \mathbf{x}_1 \underbrace{\mathbb{E}\mathbb{1}_{\mathbf{X}=\mathbf{x}_1}}_{p(\mathbf{x}_1)} + \ldots + \mathbf{x}_n \underbrace{\mathbb{E}\mathbb{1}_{\mathbf{X}=\mathbf{x}_n}}_{p(\mathbf{x}_n)}\end{aligned}$$

by property (i) of the definition of expectation. ♦

**Definition (Empirical Distribution)** The **empirical distribution** for data $x_1, \ldots, x_n$ assigns to each singleton $\{x\}$ a probability equal to its number of occurrences in the data divided by $n$, that is $\mathbb{P}\{x\} = \frac{\sum_i \mathbb{1}_{x_i=x}}{n}$.

Empirical distributions constitute an important example of discrete random vectors with finitely many possible values. If the distribution of $\mathbf{X}$ is the empirical distribution of $(\mathbf{x}_1, \ldots, \mathbf{x}_n)$, its expectation simplifies to

$$\mathbb{E}\mathbf{X} = \tfrac{1}{n} \sum_i \mathbf{x}_i$$

which we also called the *empirical mean* in Sect. 1.15. Additionally, if $\mathbf{x}_1, \ldots, \mathbf{x}_n \in \mathbb{R}$, their *empirical variance* is the variance of $\mathbf{X}$.

To keep things mathematically precise, it is good to realize that you cannot take the existence of finite expectations for granted. That is why many of our exercises have an assumption (whether explicitly or implicitly) that the expectation is *well-defined*. For random variables, expectations of $\infty$ or $-\infty$ are still considered well-defined. For other random vectors, however, the expectation has to be in the vector space to be well-defined.

**Exercise 3.11** Let $X$ be a discrete random variable whose possible values are the positive integers. In particular, suppose that $\mathbb{P}\{X = k\}$ is proportional to $1/k^2$ for $k \in \{1, 2, \ldots\}$. What is the expectation of $X$?

*Solution* Recall that $\sum_{k=1}^{\infty} \frac{1}{k^2} = \pi^2/6$, so this distribution is well-defined. However, its expectation is

$$\begin{aligned} \mathbb{E}X &= \sum_{k=1}^{\infty} k\mathbb{P}\{X = k\} \\ &= \sum_{k=1}^{\infty} k \frac{6}{\pi^2} \frac{1}{k^2} \\ &= \frac{6}{\pi^2} \sum_{k=1}^{\infty} \frac{1}{k} \\ &= \infty. \end{aligned}$$

♦

**Definition (Uncorrelated)** Random variables $X$ and $Y$ are called **uncorrelated** if $\mathbb{E}(X - \mathbb{E}X)(Y - \mathbb{E}Y) = 0$.

Exercise 3.12 Suppose $X$ is uncorrelated with each of $Y_1, \ldots, Y_n$. Show that $X$ is also uncorrelated with $a_1 Y_1 + \ldots + a_n Y_n$.

**Definition (Independence)** Two random elements **X** and **Y** are **independent** if

$$\mathbb{E}f(\mathbf{X})g(\mathbf{X}) = [\mathbb{E}f(\mathbf{X})][\mathbb{E}g(\mathbf{X})]$$

for every real-valued functions $f$ and $g$ for which $\mathbb{E}f(\mathbf{X})g(\mathbf{X})$ is finite.

Exercise 3.13 If events $E_1, \ldots, E_m$ are *independent* (meaning that their indicator functions are independent random variables) and each has probability $q$, what is the probability that at least one of them occurs?

Notice that if **X** and **Y** are independent of each other, then $f(\mathbf{X})$ and $g(\mathbf{Y})$ must also be independent of each other, for any functions $f$ and $g$. Notice also that if $X$ and $Y$ are independent, then they must also be uncorrelated.

If $X_1, \ldots, X_n$ are independent random variables with pdfs $p_1, \ldots, p_n$, then the random vector $(X_1, \ldots, X_n)$ has pdf $\prod_i p_i$ (see Sec 4.4 of Pollard (2002)). In Sect. 5.1, this fact will be used to derive the pdf of the standard multivariate Normal distribution, the one and only pdf that this text needs.

If a set of random elements are all independent of each other and all have the same distribution, they are called *independent and identically distributed* which is abbreviated *iid*.

## 3.3 Bias-Variance Decomposition

**Exercise 3.14** Let $\mathbf{X}$ be a random vector and $\mathbf{v}$ be a non-random vector. Show that if $\mathbb{E}\langle \mathbf{X}, \mathbf{v}\rangle$ is real, then it is equal to $\mathbb{E}\langle \mathbf{v}, \mathbf{X}\rangle$.

**Theorem 3.1 (Bias-Variance Decomposition)** *Let* $\mathbf{Y}$ *be a random vector with expectation* $\boldsymbol{\mu}$*, and let* $\mathbf{v}$ *be a non-random vector. The expected squared distance from* $\mathbf{Y}$ *to* $\mathbf{v}$ *decomposes as*

$$\mathbb{E}\|\mathbf{Y}-\mathbf{v}\|^2 = \|\mathbf{v}-\boldsymbol{\mu}\|^2 + \mathbb{E}\|\mathbf{Y}-\boldsymbol{\mu}\|^2.$$

*Proof* Adding and subtracting $\boldsymbol{\mu}$, then writing the norm in inner product form,

$$\begin{aligned}
\mathbb{E}\|\mathbf{Y}-\mathbf{v}\|^2 &= \mathbb{E}\|(\mathbf{Y}-\boldsymbol{\mu})-(\mathbf{v}-\boldsymbol{\mu})\|^2 \\
&= \mathbb{E}\langle(\mathbf{Y}-\boldsymbol{\mu})-(\mathbf{v}-\boldsymbol{\mu}), (\mathbf{Y}-\boldsymbol{\mu})-(\mathbf{v}-\boldsymbol{\mu})\rangle \\
&= \mathbb{E}[\langle\mathbf{Y}-\boldsymbol{\mu}, \mathbf{Y}-\boldsymbol{\mu}\rangle - \langle\mathbf{Y}-\boldsymbol{\mu}, \mathbf{v}-\boldsymbol{\mu}\rangle - \langle\mathbf{v}-\boldsymbol{\mu}, \mathbf{Y}-\boldsymbol{\mu}\rangle + \langle\mathbf{v}-\boldsymbol{\mu}, \mathbf{v}-\boldsymbol{\mu}\rangle] \\
&= \mathbb{E}\|\mathbf{Y}-\boldsymbol{\mu}\|^2 - \underbrace{\langle\mathbb{E}(\mathbf{Y}-\boldsymbol{\mu}), \mathbf{v}-\boldsymbol{\mu}\rangle}_{\mathbf{0}} - \underbrace{\mathbb{E}\langle\mathbf{v}-\boldsymbol{\mu}, \mathbf{Y}-\boldsymbol{\mu}\rangle}_{0} + \mathbb{E}\underbrace{\|\mathbf{v}-\boldsymbol{\mu}\|^2}_{\text{non-random}} \\
&= \mathbb{E}\|\mathbf{Y}-\boldsymbol{\mu}\|^2 + \|\mathbf{v}-\boldsymbol{\mu}\|^2.
\end{aligned}$$

We refer to Exercise 3.14 to see that $\langle\mathbf{v}-\boldsymbol{\mu}, \mathbf{Y}-\boldsymbol{\mu}\rangle$ has the same expectation as $\langle\mathbf{Y}-\boldsymbol{\mu}, \mathbf{v}-\boldsymbol{\mu}\rangle$. ■

The representation in Theorem 3.1 is called a *bias-variance decomposition*, terminology that comes from thinking of $\mathbf{Y}$ as an estimator for $\mathbf{v}$. The first term is a generalization of the *variance* of $\mathbf{Y}$, while the second term is a generalization of the *squared bias* of $\mathbf{Y}$ for $\mathbf{v}$ (Fig. 3.1).

**Exercise 3.15** Let $\mathbf{Y}$ be a random vector with expectation $\boldsymbol{\mu}$. Find the non-random vector $\mathbf{v}$ that minimizes $\mathbb{E}\|\mathbf{Y}-\mathbf{v}\|^2$.

*Solution* By the bias-variance decomposition, the objective function equals $\|\mathbf{v}-\boldsymbol{\mu}\|^2 + \mathbb{E}\|\mathbf{Y}-\boldsymbol{\mu}\|^2$. The second term does not depend on $\mathbf{v}$, so we can minimize the sum by taking $\mathbf{v}$ to be $\boldsymbol{\mu}$ which makes the first term zero. ♦

**Exercise 3.16** Explain how Exercise 2.2 is an instance of the bias-variance decomposition.

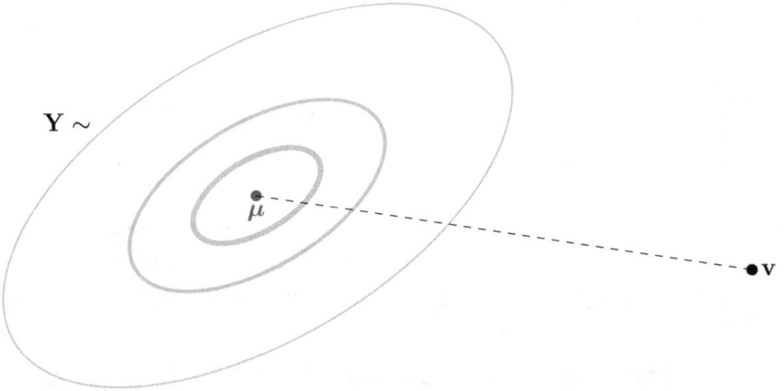


**Fig. 3.1** The bias-variance decomposition (Theorem 3.1) represents the expected squared distance from $\mathbf{Y}$ to $\mathbf{v}$ as the sum of the expected squared distance from $\mathbf{Y}$ to its expectation $\boldsymbol{\mu}$ plus the squared distance from $\boldsymbol{\mu}$ to $\mathbf{v}$

*Solution* If the distribution of the random variable $Y$ is the empirical distribution defined by $\mathbf{y} = (y_1, \ldots, y_n)$, then its expectation is $\bar{y}$. By the bias-variance decomposition,

$$\mathbb{E}(Y - a)^2 = (a - \mathbb{E}Y)^2 + \mathbb{E}(Y - \mathbb{E}Y)^2$$

$$\Updownarrow$$

$$\frac{1}{n}\sum_i (y_i - a)^2 = (a - \bar{y})^2 + \frac{1}{n}\sum_i (y_i - \bar{y})^2.$$

♦

**Exercise 3.17** Let $\mathbf{Y}$ be a random vector that is an *unbiased estimator* for $\boldsymbol{\theta}$, that is $\mathbb{E}\mathbf{Y} = \boldsymbol{\theta}$. If $\lambda \in \mathbb{R}$, express $\|\mathbb{E}(\lambda \mathbf{Y}) - \boldsymbol{\theta}\|^2$ (which can be thought of as the *squared bias* of the estimator $\lambda \mathbf{Y}$) in terms of $\lambda$ and $\|\boldsymbol{\theta}\|^2$.

*Solution*

$$\begin{aligned} \|\mathbb{E}(\lambda \mathbf{Y}) - \boldsymbol{\theta}\|^2 &= \|\lambda \underbrace{\mathbb{E}\mathbf{Y}}_{\boldsymbol{\theta}} - \boldsymbol{\theta}\|^2 \\ &= \|(\lambda - 1)\boldsymbol{\theta}\|^2 \\ &= (1 - \lambda)^2 \|\boldsymbol{\theta}\|^2 \end{aligned}$$

Note that the factor $(\lambda - 1)^2$ is equal to $(1 - \lambda)^2$ which is a bit more intuitive when $\lambda \in [0, 1]$. ♦

**Exercise 3.18** Let $\mathbf{Y}$ be a random vector, and let $\lambda \in \mathbb{R}$. Express $\mathbb{E}\|\lambda \mathbf{Y} - \mathbb{E}(\lambda \mathbf{Y})\|^2$ in terms of $\lambda$ and $\mathbb{E}\|\mathbf{Y} - \mathbb{E}\mathbf{Y}\|^2$.

*Solution* Factoring out $\lambda$,

$$\begin{aligned}\mathbb{E}\|\lambda\mathbf{Y} - \mathbb{E}(\lambda\mathbf{Y})\|^2 &= \mathbb{E}\|\lambda(\mathbf{Y} - \mathbb{E}\mathbf{Y})\|^2 \\ &= \lambda^2\mathbb{E}\|\mathbf{Y} - \mathbb{E}\mathbf{Y}\|^2.\end{aligned}$$

♦

**Exercise 3.19** Let $\mathbf{Y}$ be a random vector that is an *unbiased estimator* for $\boldsymbol{\theta} \in \mathbb{R}^n$. Use the bias-variance decomposition along with your results from Exercises 3.17 and 3.18 to find an expression for $\lambda \in \mathbb{R}$ (in terms of $\|\boldsymbol{\theta}\|^2$ and $\mathbb{E}\|\mathbf{Y} - \boldsymbol{\theta}\|^2$) for which $\mathbb{E}\|\boldsymbol{\theta} - \lambda\mathbf{Y}\|^2$ is as small as possible.

*Solution* By the bias-variance decomposition and our previous results,

$$\begin{aligned}\mathbb{E}\|\boldsymbol{\theta} - \lambda\mathbf{Y}\|^2 &= \|\mathbb{E}(\lambda\mathbf{Y}) - \boldsymbol{\theta}\|^2 + \mathbb{E}\|\lambda\mathbf{Y} - \mathbb{E}(\lambda\mathbf{Y})\|^2 \\ &= (1-\lambda)^2\|\boldsymbol{\theta}\|^2 + \lambda^2\mathbb{E}\|\mathbf{Y} - \mathbb{E}\mathbf{Y}\|^2.\end{aligned}$$

Taking the derivative with respect to $\lambda$, and setting it to zero, we get the critical $\lambda^*$:

$$(1-\lambda^*)\|\boldsymbol{\theta}\|^2 = \lambda^*\mathbb{E}\|\mathbf{Y} - \mathbb{E}\mathbf{Y}\|^2$$

is solved by $\lambda^* = \frac{\|\boldsymbol{\theta}\|^2}{\|\boldsymbol{\theta}\|^2 + \mathbb{E}\|\mathbf{Y} - \mathbb{E}\mathbf{Y}\|^2}$. Realize of course that when estimating an unknown parameter $\boldsymbol{\theta}$, we cannot actually calculate this optimal value. ♦

## 3.4 Covariance

**Definition (Random Matrix)** A **random matrix** is a rectangular arrangement of random variables. If $\mathbb{X}$ is a random matrix, its expectation $\mathbb{E}\mathbb{X}$ is defined to be the matrix whose $(i, j)$ entry is the expectation of the random variable in the $(i, j)$ entry of $\mathbb{X}$.

The operations on ordinary real matrices are naturally extended to random matrices.

**Exercise 3.20** Let $\mathbb{X}$ be a random matrix whose entries have finite expectations, and let $\mathbb{M}$ be a non-random matrix. Assuming $\mathbb{M}\mathbb{X}$ is well-defined, show that $\mathbb{E}\mathbb{M}\mathbb{X} = \mathbb{M}\mathbb{E}\mathbb{X}$. Alternatively, assuming $\mathbb{X}\mathbb{M}$ is well-defined, show that $\mathbb{E}\mathbb{X}\mathbb{M} = (\mathbb{E}\mathbb{X})\mathbb{M}$.

**Definition (Covariance)** Let $\mathbf{Y} = (Y_1, \ldots, Y_n)$ be an $\mathbb{R}^n$-valued random vector. The **covariance matrix** of $\mathbf{Y}$ is the $n \times n$ matrix whose $(i, j)$ entry equals the **covariance** between $Y_i$ and $Y_j$ which is defined to be $\mathbb{E}[(Y_i - \mathbb{E}Y_i)(Y_j - \mathbb{E}Y_j)]$. The **variance** of a random variable $Y$ is its expected squared deviation from its expectation $\mathbb{E}(Y - \mathbb{E}Y)^2$. The **standard deviation** is the square root of the variance. The **correlation matrix** of $\mathbf{Y}$ is the $n \times n$ matrix whose $(i, j)$ entry equals the **correlation** between $Y_i$ and $Y_j$ which is defined to be their covariance divided by the product of their standard deviations.

Notice from the definition that covariance matrices must be symmetric. Notice also that the diagonals are the variances of the coordinate random variables.

**Exercise 3.21** Show that an alternative expression for the covariance matrix of $\mathbf{Y}$ is $\mathbb{E}[(\mathbf{Y} - \mathbb{E}\mathbf{Y})(\mathbf{Y} - \mathbb{E}\mathbf{Y})^T]$.

**Exercise 3.22** Let $\mathbf{Y}$ be an $\mathbb{R}^n$-valued random vector, and let $\mathbf{v} \in \mathbb{R}^n$. Use Exercise 3.21 to show that the covariance of $\mathbf{v} + \mathbf{Y}$ has the same covariance matrix as $\mathbf{Y}$.

*Solution*

$$\begin{aligned}\operatorname{cov}(\mathbf{v} + \mathbf{Y}) &= \mathbb{E}[(\mathbf{v} + \mathbf{Y} - \mathbb{E}(\mathbf{v} + \mathbf{Y}))(\mathbf{v} + \mathbf{Y} - \mathbb{E}(\mathbf{v} + \mathbf{Y}))^T] \\ &= \mathbb{E}[(\mathbf{Y} - \mathbb{E}\mathbf{Y})(\mathbf{Y} - \mathbb{E}\mathbf{Y})^T] \\ &= \operatorname{cov}\mathbf{Y}\end{aligned}$$

♦

**Exercise 3.23** Let $\mathbf{Y}$ be a random vector with covariance matrix $\mathbb{C}$. Let $\mathbf{v}$ be a non-random vector, and let $\mathbb{M}$ be a real matrix. Show that the covariance of $\mathbf{v} + \mathbb{M}\mathbf{Y}$ is $\mathbb{M}\mathbb{C}\mathbb{M}^T$.

*Solution* By Exercise 3.22, $\operatorname{cov}(\mathbf{v} + \mathbb{M}\mathbf{Y}) = \operatorname{cov}\mathbb{M}\mathbf{Y}$.

$$\begin{aligned}\operatorname{cov}\mathbb{M}\mathbf{Y} &= \mathbb{E}[(\mathbb{M}\mathbf{Y} - \mathbb{E}\mathbb{M}\mathbf{Y})(\mathbb{M}\mathbf{Y} - \mathbb{E}\mathbb{M}\mathbf{Y})^T] \\ &= \mathbb{E}[(\mathbb{M}\mathbf{Y} - \mathbb{M}\mathbb{E}\mathbf{Y})(\mathbb{M}\mathbf{Y} - \mathbb{M}\mathbb{E}\mathbf{Y})^T] \\ &= \mathbb{E}[(\mathbb{M}(\mathbf{Y} - \mathbb{E}\mathbf{Y}))(\mathbb{M}(\mathbf{Y} - \mathbb{E}\mathbf{Y}))^T] \\ &= \mathbb{M}(\mathbb{E}[(\mathbf{Y} - \mathbb{E}\mathbf{Y})(\mathbf{Y} - \mathbb{E}\mathbf{Y})^T])\mathbb{M}^T \\ &= \mathbb{M}\mathbb{C}\mathbb{M}^T\end{aligned}$$

♦

**Exercise 3.24** Show that every covariance matrix is positive semi-definite.

*Solution* To satisfy the definition, we need to show that every quadratic form is non-negative. We'll use the covariance expression from Exercise 3.21 and consider its quadratic form for an arbitrary vector $\mathbf{v}$,

$$\begin{aligned}\mathbf{v}^T\mathbb{E}[(\mathbf{Y} - \mathbb{E}\mathbf{Y})(\mathbf{Y} - \mathbb{E}\mathbf{Y})^T]\mathbf{v} &= \mathbb{E}[\mathbf{v}^T(\mathbf{Y} - \mathbb{E}\mathbf{Y})(\mathbf{Y} - \mathbb{E}\mathbf{Y})^T]\mathbf{v} \\ &= \mathbb{E}[\mathbf{v}^T(\mathbf{Y} - \mathbb{E}\mathbf{Y})(\mathbf{Y} - \mathbb{E}\mathbf{Y})^T\mathbf{v}] \\ &= \mathbb{E}\langle\mathbf{Y} - \mathbb{E}\mathbf{Y}, \mathbf{v}\rangle^2.\end{aligned}$$

The expectation of a non-negative random variable has to be non-negative. ♦

**Exercise 3.25** Show that $\mathbb{E}\|\mathbf{X} - \mathbb{E}\mathbf{X}\|^2 = \text{tr}\,(\text{cov}\,\mathbf{X})$.

*Solution*

$$\begin{aligned}\mathbb{E}\|\mathbf{X} - \mathbb{E}\mathbf{X}\|^2 &= \mathbb{E}[(X_1 - \mathbb{E}X_1)^2 + \ldots + (X_n - \mathbb{E}X_n)^2] \\ &= \mathbb{E}(X_1 - \mathbb{E}X_1)^2 + \ldots + \mathbb{E}(X_n - \mathbb{E}X_n)^2\end{aligned}$$

These variances are the diagonals of the covariance matrix, so its trace is their sum. ♦

**Exercise 3.26** Suppose $X_1, \ldots, X_n$ are uncorrelated random variables. Show that the variance of their sum equals the sum of their variances.

**Exercise 3.27** Let $\boldsymbol{\epsilon}$ be a random vector with expectation $\mathbf{0}$ and covariance matrix $\sigma^2\mathbb{I}$. Let $\mathbf{v}$ be a non-random vector, and let $\mathbb{H}$ be an orthogonal projection matrix. Find the covariance matrix of $\mathbb{H}(\mathbf{v} + \boldsymbol{\epsilon})$.

*Solution* Distribute the matrix multiplication to get $\mathbb{H}\mathbf{v} + \mathbb{H}\boldsymbol{\epsilon}$. According to Exercise 3.23, the covariance is

$$\begin{aligned}\mathbb{H}(\sigma^2\mathbb{I})\mathbb{H}^T &= \sigma^2\mathbb{H}\mathbb{H}^T \\ &= \sigma^2\mathbb{H}\end{aligned}$$

by symmetry and idempotence of orthogonal projection matrices. ♦

**Exercise 3.28** Let $\mathbf{X}$ have expectation $\boldsymbol{\mu}_{\mathbf{X}}$ and $\mathbf{Y}$ have expectation $\boldsymbol{\mu}_{\mathbf{Y}}$. Show that the expected inner product between the centered vectors $\mathbf{X} - \boldsymbol{\mu}_{\mathbf{X}}$ and $\mathbf{Y} - \boldsymbol{\mu}_{\mathbf{Y}}$ is the same as the expected inner product when only one of them is centered.

*Solution*

$$\begin{aligned}\mathbb{E}\langle \mathbf{X} - \boldsymbol{\mu}_{\mathbf{X}}, \mathbf{Y} - \boldsymbol{\mu}_{\mathbf{Y}} \rangle &= \mathbb{E}\langle \mathbf{X} - \boldsymbol{\mu}_{\mathbf{X}}, \mathbf{Y} \rangle - \mathbb{E}\langle \mathbf{X} - \boldsymbol{\mu}_{\mathbf{X}}, \boldsymbol{\mu}_{\mathbf{Y}} \rangle \\ &= \mathbb{E}\langle \mathbf{X} - \boldsymbol{\mu}_{\mathbf{X}}, \mathbf{Y} \rangle - \langle \underbrace{\mathbb{E}\mathbf{X} - \boldsymbol{\mu}_{\mathbf{X}}}_{\mathbf{0}}, \boldsymbol{\mu}_{\mathbf{Y}} \rangle \\ &= \mathbb{E}\langle \mathbf{X} - \boldsymbol{\mu}_{\mathbf{X}}, \mathbf{Y} \rangle\end{aligned}$$

The same argument works for $\mathbf{Y} - \boldsymbol{\mu}_{\mathbf{Y}}$ if you keep Exercise 3.14 in mind. ♦

**Exercise 3.29** Use Exercise 3.28 to observe that

$$\langle x - \bar{x}\mathbf{1}, y \rangle = \langle x - \bar{x}\mathbf{1}, y - \bar{y}\mathbf{1} \rangle.$$

*Solution* Let the *joint* distribution of $(X, Y)$ be the empirical distribution of $(x_1, y_1), \ldots, (x_n, y_n)$.

$$\begin{aligned}
\langle \mathbf{x} - \bar{x}\mathbf{1}, \mathbf{y} \rangle &= n\frac{1}{n}\sum_i [(x_i - \bar{x})y_i] \\
&= n\mathbb{E}[(X - \mathbb{E}X)Y] \\
&= n\mathbb{E}[(X - \mathbb{E}X)(Y - \mathbb{E}Y)] \\
&= n\frac{1}{n}\sum_i [(x_i - \bar{x})(y_i - \bar{y})] \\
&= \langle \mathbf{x} - \bar{x}\mathbf{1}, \mathbf{y} - \bar{y}\mathbf{1} \rangle
\end{aligned}$$

♦

**Exercise 3.30** Let $\mathbf{X}$ be a random vector mapping to a real vector space, and let $\mathbf{v}$ be a non-random vector. Show that the variance of the coordinate of $\mathbf{X}$ with respect to $\mathbf{u}$ is the same as the variance of the coordinate of $\mathbf{X} + \mathbf{v}$ with respect to $\mathbf{u}$.

*Solution* The difference between $\langle \mathbf{X} + \mathbf{v}, \mathbf{u} \rangle$ and $\langle \mathbf{X}, \mathbf{u} \rangle$ is $\langle \mathbf{v}, \mathbf{u} \rangle$ which is non-random. By Exercise 3.23, we can conclude that they must therefore have the same variance. ♦

**Exercise 3.31** If $\mathbf{X}$ has expectation $\boldsymbol{\mu}$, find the expectation of the *centered* version $\mathbf{X} - \boldsymbol{\mu}$.

*Solution*

$$\begin{aligned}
\mathbb{E}(\mathbf{X} - \boldsymbol{\mu}) &= \underbrace{\mathbb{E}\mathbf{X}}_{\boldsymbol{\mu}} - \boldsymbol{\mu} \\
&= \mathbf{0}
\end{aligned}$$

♦

If the distribution of an $\mathbb{R}^n$-valued random vector $\mathbf{X}$ is the empirical distribution of $(\mathbf{x}_1, \ldots, \mathbf{x}_n)$, then its covariance matrix is exactly the *empirical covariance matrix* of the data, as defined in Sect. 1.15. In fact, our findings in that section can be generalized to other distributions on $\mathbb{R}^n$ as we'll now discuss.

Suppose $\mathbf{X}$ has expectation $\boldsymbol{\mu}$ and covariance matrix $\mathbb{C}$; let $\lambda_1\mathbf{q}_1\mathbf{q}_1^T + \ldots + \lambda_n\mathbf{q}_n\mathbf{q}_n^T$ be a spectral decomposition of $\mathbb{C}$. Let us figure out in which direction its distribution "spreads out" the most from its expectation. Specifically, for what unit vector $\mathbf{u}$ does the coordinate of $\mathbf{X}$ have the largest variance? Based on Exercise 3.30, the variance of $\langle \mathbf{X}, \mathbf{u} \rangle$ is the same as the variance of $\langle \mathbf{X} - \boldsymbol{\mu}, \mathbf{u} \rangle$. The latter random variable has expectation zero according to Exercise 3.31, so its variance is simply its expected square. With some clever regrouping of matrix multiplications, we can bring $\mathbf{u}$ outside of the expectation.[2]

$$\mathbb{E}[(\mathbf{X}-\boldsymbol{\mu})^T\mathbf{u}]^2 = \mathbb{E}\mathbf{u}^T(\mathbf{X}-\boldsymbol{\mu})(\mathbf{X}-\boldsymbol{\mu})^T\mathbf{u}$$
$$= \mathbf{u}^T\underbrace{[\mathbb{E}(\mathbf{X}-\boldsymbol{\mu})(\mathbf{X}-\boldsymbol{\mu})^T]}_{\mathbb{C}}\mathbf{u}$$

We see that the variance is exactly the quadratic form for the covariance matrix. We know from Exercise 1.77 that it is maximized by the principal eigenvector in which case its value is the largest eigenvalue. Furthermore, the variance of the coordinate of $\mathbf{X}$ with respect to a unit eigenvector $\mathbf{q}_j$ equals the corresponding eigenvalue $\mathbf{q}_j^T\mathbb{C}\mathbf{q}_j = \lambda_j$. (By applying these results to an empirical distribution, one rediscovers the findings from Sect. 1.15.)

**Definition (Degenerate)** If a random vector's covariance matrix is not positive definite, then its distribution is called **degenerate**.

If $\mathbf{X}$ is an $\mathbb{R}^n$-valued random vector with a degenerate distribution, then all of its probability lies in a proper subspace of $\mathbb{R}^n$. Specifically, $\mathbf{X}$ inhabits the span of the covariance matrix's eigenvectors that have positive eigenvalues. We know that it cannot deviate from this subspace, because its variance (eigenvalue) is zero in any orthogonal direction.

## 3.5 Standardizing

From your previous studies, you have likely *standardized* a random variable by subtracting its expectation then dividing by its standard deviation; the resulting random variable has expectation equal to zero and standard deviation equal to one. There is a generalization of this process for random vectors. The first step is to subtract the expectation, of course, but the second step is a little more complicated: multiply by the square root[3] of the inverse of the covariance matrix.[4] The resulting random vector has expectation equal to the zero vector and covariance equal to the identity matrix as we'll see in Exercise 3.33.

**Exercise 3.32** Let $\mathbb{M}$ be a positive definite matrix. Based on Exercises 1.24 and 1.58, explain why the inverse of the square root of $\mathbb{M}$ is the same as the square root of the inverse of $\mathbb{M}$.

*Solution* To find the square root of a positive semi-definite matrix, you replace the eigenvalues by their square roots. To find the inverse of an invertible symmetric

[2]These steps are analogous to Exercise 1.81 in reverse.

[3]The square root of a matrix was defined in Exercise 1.58.

[4]Notice that standardizing requires the covariance matrix to be invertible.

matrix, you replace the eigenvalues by their reciprocals. No matter which order you do these two operations in, you end up with the same matrix:

$$\frac{1}{\sqrt{\lambda_1}}\mathbf{q}_1\mathbf{q}_1^T + \ldots + \frac{1}{\sqrt{\lambda_n}}\mathbf{q}_n\mathbf{q}_n^T,$$

where $\mathbf{q}_1, \ldots, \mathbf{q}_n$ are eigenvectors of $\mathbb{M}$ with eigenvalues $\lambda_1, \ldots, \lambda_n$. ♦

**Exercise 3.33** Let $\mathbf{Y}$ have expectation $\boldsymbol{\mu}$ and covariance matrix $\mathbb{C}$. Find the expectation and covariance of $\mathbb{C}^{-1/2}(\mathbf{Y} - \boldsymbol{\mu})$.

*Solution* The random vector $\mathbf{Y} - \boldsymbol{\mu}$ has expectation zero, so based on Exercise 3.9, $\mathbb{C}^{-1/2}(\mathbf{Y}-\boldsymbol{\mu})$ has expectation $\mathbb{C}^{-1/2}\mathbf{0} = \mathbf{0}$. For the covariance, we apply the formula from Exercise 3.23 to get

$$\begin{aligned}
\text{cov}\,[\mathbb{C}^{-1/2}(\mathbf{Y} - \boldsymbol{\mu})] &= \mathbb{C}^{-1/2}\mathbb{C}(\mathbb{C}^{-1/2})^T \\
&= \underbrace{\mathbb{C}^{-1/2}\mathbb{C}^{1/2}}_{\mathbb{I}}\,\underbrace{\mathbb{C}^{1/2}\mathbb{C}^{-1/2}}_{\mathbb{I}} \\
&= \mathbb{I}.
\end{aligned}$$

♦

The idea behind Exercise 3.33 can also be used to transform a random vector with zero expectation and identity covariance into a random vector with a specified expectation $\boldsymbol{\mu}$ and covariance $\mathbb{C}$. Solving that exercise's equation for $\mathbf{Y}$, we have

$$\mathbf{Y} = \mathbb{C}^{1/2}\mathbf{Z} + \boldsymbol{\mu}.$$

In other words, if a random vector with zero expectation and identity covariance gets multiplied by $\mathbb{C}^{1/2}$, then added with $\boldsymbol{\mu}$, the random vector resulting from that transformation has expectation $\boldsymbol{\mu}$ and covariance $\mathbb{C}$.

A concept from introductory statistics that is closely related to standardizing is the *z-score* of a number relative to a distribution; it tells you how many standard deviations above the mean that number is. We'll now introduce an important (non-negative) generalization of this concept for random vectors.

**Definition (Mahalanobis Distance)** Let $\mathbf{Y}$ be an $\mathbb{R}^n$-valued random vector with expectation $\boldsymbol{\mu}$ and a positive definite covariance matrix $\mathbb{C}$. The **Mahalanobis distance** from $\mathbf{v} \in \mathbb{R}$ to the distribution of $\mathbf{Y}$ is $\|\mathbb{C}^{-1/2}(\mathbf{v} - \boldsymbol{\mu})\|$.

**Exercise 3.34** If $\mathbf{Y}$ has expectation $\boldsymbol{\mu}$ and a positive definite covariance matrix $\mathbb{C}$, find the expected squared Mahalanobis distance from $\mathbf{Y}$ to its own distribution.

*Solution* Let $\mathbf{Z} := \mathbb{C}^{-1/2}(\mathbf{Y} - \boldsymbol{\mu})$ represent the standardized version of $\mathbf{Y}$, and let $(Z_1, \ldots, Z_n)$ represent its coordinates. Notice that the squared Mahalanobis distance from $\mathbf{Y}$ to its distribution is exactly the squared norm of the standardized version.

$$\begin{aligned}\mathbb{E}\|\mathbb{C}^{-1/2}[\mathbf{Y}-\boldsymbol{\mu}]\|^2 &= \mathbb{E}\|\mathbf{Z}\|^2 \\ &= \mathbb{E}Z_1^2 + \ldots + \mathbb{E}Z_n^2 \\ &= \underbrace{\text{var } Z_1}_{1} + \ldots + \underbrace{\text{var } Z_n}_{1} \\ &= n\end{aligned}$$

The expected squared Mahalanobis distance is the dimension of the vector space that $\mathbf{Y}$ inhabits. ♦

Squared Mahalanobis distance is also commonly expressed as a quadratic form:

$$\begin{aligned}\|\mathbb{C}^{-1/2}(\mathbf{v}-\boldsymbol{\mu})\|^2 &= [\mathbb{C}^{-1/2}(\mathbf{v}-\boldsymbol{\mu})]^T[\mathbb{C}^{-1/2}(\mathbf{v}-\boldsymbol{\mu})] \\ &= (\mathbf{v}-\boldsymbol{\mu})^T(\mathbb{C}^{-1/2})^T\mathbb{C}^{-1/2}(\mathbf{v}-\boldsymbol{\mu}) \\ &= (\mathbf{v}-\boldsymbol{\mu})^T\mathbb{C}^{-1}(\mathbf{v}-\boldsymbol{\mu}).\end{aligned}$$

The Mahalanobis distance from a distribution to its mean is zero, it increases monotonically as you move away from the mean in any direction, and its level curves are shaped as *ellipsoids* with the covariance eigenvector directions as axes.[5]

We can gain a great deal of intuition about Mahalanobis distance by once again thinking about spectral decompositions. Suppose $\mathbb{C}$ has orthonormal eigenvectors $\mathbf{q}_1, \ldots, \mathbf{q}_n$ with positive eigenvalues $\lambda_1, \ldots, \lambda_n$. The Mahalanobis distance from $\boldsymbol{\mu}$ to the distribution is 0. Consider "traveling" from $\boldsymbol{\mu}$ in the direction of a unit vector $\mathbf{u}$ by a distance of $\alpha \geq 0$. What is the squared Mahalanobis distance of the point where you landed $\boldsymbol{\mu} + \alpha\mathbf{u}$? We can invoke Exercise 1.76 to see that its proportional to a linear combination of the eigenvalues of $\mathbb{C}^{-1}$.

$$\begin{aligned}[(\boldsymbol{\mu}+\alpha\mathbf{u})-\boldsymbol{\mu}]^T\mathbb{C}^{-1}[(\boldsymbol{\mu}+\alpha\mathbf{u})-\boldsymbol{\mu}] &= \alpha^2\mathbf{u}^T\mathbb{C}^{-1}\mathbf{u} \\ &= \alpha^2\left(\langle\mathbf{u},\mathbf{q}_1\rangle^2\lambda_1^{-1} + \ldots + \langle\mathbf{u},\mathbf{q}_n\rangle^2\lambda_n^{-1}\right)\end{aligned}$$

We see that the Mahalanobis distance increases most slowly in the direction of the covariance matrix's principal eigenvector, and it increases most rapidly in the direction of the eigenvector of $\mathbb{C}$ with the smallest eigenvalue.

This quadratic form is particularly simple if you travel in one of the eigenvector directions. The Mahalanobis distance from $\alpha\mathbf{q}_j$ to the distribution is $\frac{\alpha}{\sqrt{\lambda_j}}$. From Sect. 3.4, we know that the standard deviation in the direction of $\mathbf{q}_j$ is $\sqrt{\lambda_j}$. Putting these two observations together provides us with a neat statement: a point that is $z$

[5]The reader can consult Grötschel et al. (1988) (Sec 3.1) or another source for the definition of ellipsoids and an explanation of their shapes.

standard deviations from $\boldsymbol{\mu}$ in the direction of $\mathbf{q}_j$ has Mahalanobis distance $z$ to the distribution.

We can also see a natural way to extend this section's ideas to the case of degenerate distributions. Such a distribution's covariance matrix is not invertible, but one can instead use the Moore–Penrose inverse $\mathbb{C}^-$ (Sect. 1.13) which behaves like an inverse in the column space of $\mathbb{C}$. For standardizing, we'll need to make use of the Moore–Penrose inverse of $\mathbb{C}^{1/2}$, which is the same as the square root of $\mathbb{C}^-$ (for the same reason given in Exercise 3.32). Let us generalize our previous notation, and proceed with the convention that $\mathbb{C}^{-1/2}$ should be interpreted as the Moore–Penrose inverse of the square root of $\mathbb{C}$. If $\mathbf{Y}$ has expectation $\boldsymbol{\mu}$ and its covariance $\mathbb{C}$ has rank $d$, then $(\mathbb{C}^{1/2})^-(\mathbf{Y}-\boldsymbol{\mu})$ is a random vector in $\mathbb{R}^d$ with expectation $\mathbf{0} \in \mathbb{R}^d$ and covariance $(\mathbb{C}^{-1/2})\mathbb{C}(\mathbb{C}^{-1/2}) = \mathbb{I}_d$. (Use spectral decompositions to see why this product of matrices is the identity.)

## 3.6 Expectation of Quadratic Form

A clever trick for manipulating quadratic forms comes from realizing that the result of the matrix multiplications is just a number, which is a $1 \times 1$ matrix and is equal to its own trace; this allows us to make use of the cyclic permutation property of the trace operator (Exercise 1.51). If $\mathbf{Y}$ is a random vector *with expectation* $\mathbf{0}$ and covariance matrix $\mathbb{C}$,

$$\begin{aligned}\mathbb{E}\mathbf{Y}^T\mathbb{M}\mathbf{Y} &= \mathbb{E}\,\mathrm{tr}\,(\mathbf{Y}^T\mathbb{M}\mathbf{Y})\\ &= \mathbb{E}\,\mathrm{tr}\,(\mathbb{M}\mathbf{Y}\mathbf{Y}^T)\\ &= \mathrm{tr}\,(\mathbb{M}\,\mathbb{E}\mathbf{Y}\mathbf{Y}^T)\\ &= \mathrm{tr}\,[\mathbb{M}\mathbb{C}].\end{aligned}$$

Note that the trace operator commutes with the expectation operator because you get the same result if you take the expectations of the diagonals of a matrix before summing them or if you sum them before taking the expectation.

Now let us drop the mean-zero assumption, and let $\mathbf{Y}$ have an arbitrary expectation $\boldsymbol{\mu}$. The centered version $\mathbf{Y}-\boldsymbol{\mu}$ has the same covariance matrix that $\mathbf{Y}$ does (Exercise 3.23), and because $\mathbf{Y}-\boldsymbol{\mu}$ has expectation $\mathbf{0}$, we can apply the above formula to $\mathbb{E}(\mathbf{Y}-\boldsymbol{\mu})^T\mathbb{M}(\mathbf{Y}-\boldsymbol{\mu})$.

$$\begin{aligned}\mathbb{E}\mathbf{Y}^T\mathbb{M}\mathbf{Y} &= \mathbb{E}(\mathbf{Y}-\boldsymbol{\mu}+\boldsymbol{\mu})^T\mathbb{M}(\mathbf{Y}-\boldsymbol{\mu}+\boldsymbol{\mu})\\ &= \mathbb{E}(\mathbf{Y}-\boldsymbol{\mu})^T\mathbb{M}(\mathbf{Y}-\boldsymbol{\mu}) + \underbrace{\mathbb{E}(\mathbf{Y}-\boldsymbol{\mu})^T\mathbb{M}\boldsymbol{\mu}}_{0} + \underbrace{\mathbb{E}\boldsymbol{\mu}^T\mathbb{M}(\mathbf{Y}-\boldsymbol{\mu})}_{0} + \mathbb{E}\boldsymbol{\mu}^T\mathbb{M}\boldsymbol{\mu}\\ &= \mathrm{tr}\,[\mathbb{M}\mathbb{C}] + \boldsymbol{\mu}^T\mathbb{M}\boldsymbol{\mu}\end{aligned}$$

If $\mathbf{Y}$ is $\mathbb{R}^n$-valued and $\mathbb{M}$ is a real symmetric matrix, then this formula can also be expressed in terms of norms as

$$\mathbb{E}\|\mathbb{M}^{1/2}\mathbf{Y}\|^2 = \operatorname{tr}[\mathbb{M}\mathbb{C}] + \|\mathbb{M}^{1/2}\boldsymbol{\mu}\|^2.$$

More specifically, with an orthogonal projection matrix $\mathbb{H} \in \mathbb{R}^{n\times n}$,

$$\mathbb{E}\|\mathbb{H}\mathbf{Y}\|^2 = \operatorname{tr}[\mathbb{H}\mathbb{C}] + \|\mathbb{H}\boldsymbol{\mu}\|^2 \tag{3.1}$$

because $\mathbb{H}$ is its own square root. (See also Exercise 1.80.)

**Exercise 3.35** Let $\mathbb{H}$ be the orthogonal projection matrix onto a $d$-dimensional subspace $\mathcal{S} \subseteq \mathbb{R}^n$, and let $\mathbf{Y}$ be a random vector with covariance matrix $\sigma^2\mathbb{I}$. Show that

$$\mathbb{E}\|\mathbb{H}\mathbf{Y}\|^2 = d\sigma^2 + \|\mathbb{H}\boldsymbol{\mu}\|^2.$$

*Solution* By comparison to Eq. 3.1, all that remains is to verify that the trace of $\mathbb{H}\sigma^2\mathbb{I}$ is $d\sigma^2$.

$$\begin{aligned}\operatorname{tr}[\mathbb{H}\sigma^2\mathbb{I}] &= \sigma^2 \operatorname{tr}\mathbb{H}\\ &= d\sigma^2\end{aligned}$$

because according to Exercise 1.67 the trace of an orthogonal projection matrix equals the dimension of the subspace that it projects onto. ♦

Notice that if $\boldsymbol{\mu}$ is orthogonal to the subspace that $\mathbb{H}$ projects onto, then the second term on the right vanishes. As another special case of interest, if $\boldsymbol{\mu}$ is in the subspace, then the second term simplifies to $\|\boldsymbol{\mu}\|^2$.

**Exercise 3.36** Let $X_1, \ldots, X_n$ all have expectation $\boldsymbol{\mu}_X$, and let $Y_1, \ldots, Y_n$ all have expectation $\boldsymbol{\mu}_Y$. Suppose $\operatorname{cov}(X_i, Y_j)$ equals $\sigma_{X,Y}$ if $i = j$ and zero otherwise. Find the expectation of

$$\sum_i (X_i - \bar{X})(Y_i - \bar{Y}).$$

Now that we understand random vectors, we are ready to use them to *finally* define linear models in Chap. 4.

# Chapter 4
# Linear Models

In Chap. 2, we learned how to do least-squares regression when the set of possible prediction vectors comprise a subspace: the least-squares prediction vector is the orthogonal projection of the data onto that subspace. Now we will formulate probabilistic models for which the possible expectation vectors of the response variable comprise a subspace. Each type of model we discuss has a counterpart in our earlier study of regression, and we will reanalyze the least-squares predictions and coefficients in the context of the model. To make it easy for the reader to compare the material from the two chapters, their structures perfectly parallel each other.

## 4.1 Modeling Data

A model is a mathematical formulation that is proposed to approximately represent a real-world phenomenon. The model is a *simplified* version of reality that is typically imperfect but may still be useful.[1] Often the model is a set of statements that are all under consideration, and the task remains to select one statement in particular based on your observations of real-world data. Many real-world quantities cannot be predicted exactly, either due to limitations in our knowledge or due to the nature of physical reality. In those cases, we use random vectors in the formulations and call them *probabilistic* (as opposed to *deterministic*) models. For us, the response variable values will be modeled by a random vector $\mathbf{Y}$ whose expectation is a function of non-random explanatory variables.

Recall that least-squares linear regression uses the orthogonal projection of the response vector $\mathbf{Y}$ onto a subspace $\mathcal{S}$ as the prediction vector $\widehat{\mathbf{Y}}$. Let us consider $\widehat{\mathbf{Y}}$ to be an *estimator* for $\mathbb{E}\mathbf{Y}$. Realize that $\mathbb{E}\widehat{\mathbf{Y}}$ equals the orthogonal projection of $\mathbb{E}\mathbf{Y}$ onto $\mathcal{S}$. Thus if $\mathbb{E}\mathbf{Y}$ is in $\mathcal{S}$, then $\widehat{\mathbf{Y}}$ is an unbiased estimator for it.

[1] In general, there is a trade-off between simplicity and accuracy.

 
W. D. Brinda, *Visualizing Linear Models*,
https://doi.org/10.1007/978-3-030-64167-2_4

To think about how *good* an estimator is, it is often useful to know its covariance matrix. Let us suppose that $Y_1, \ldots, Y_n$ are uncorrelated and all have the same variance $\sigma^2$. Then by Exercise 3.27 the covariance matrix of $\widehat{\mathbf{Y}}$ is $\sigma^2$ times the orthogonal projection matrix onto $\mathcal{S}$. This matrix can be expressed as $\mathbb{M}(\mathbb{M}^T\mathbb{M})^-\mathbb{M}$ if $\mathbb{M}$ is the design matrix used for the regression.

We will also want to *consider the least-squares coefficients as estimators for the true coefficients* when the expectation of $\mathbf{Y}$ is indeed a linear function of coefficients. In this chapter, we will analyze these estimators and related quantities without assuming a particular distribution for the *errors*. In Chap. 6, we will assume specifically that the errors are Normal, which will allow us to devise hypothesis tests and confidence intervals.

As in Chap. 2, we will start with the simplest case, then each subsection will generalize the one that came before it. And as before, we will draw both the *observations* pictures and the *variables* pictures this time with a few extra ingredients that come from the models.

### *4.1.1 Location Models*

**Definition (Location Model)** Let $Y_1, \ldots, Y_n$ be random variables representing a *response* variable. A location model assumes that each observation can be represented by

$$Y_i = \alpha + \epsilon_i,$$

where $\alpha \in \mathbb{R}$ and $\epsilon_1, \ldots, \epsilon_n$ are random variables that each have expectation 0.

In this type of model, the expectation of every response value is $\alpha$. The random variables $\epsilon_1, \ldots, \epsilon_n$ are called the *errors*; they are the deviations of the response values from their expectations.

As in Chap. 2, the key to our analyses will be translating statements about observations into statements about variable vectors. To this end, the location model assumption can be expressed for $\mathbf{Y} = (Y_1, \ldots, Y_n)$ as

$$\begin{bmatrix} Y_1 \\ \vdots \\ Y_n \end{bmatrix} = \begin{bmatrix} \alpha + \epsilon_1 \\ \vdots \\ \alpha + \epsilon_n \end{bmatrix}$$
$$= \alpha\mathbf{1} + \boldsymbol{\epsilon},$$

where $\boldsymbol{\epsilon} := (\epsilon_1, \ldots, \epsilon_n)$ is a random vector with expectation $\mathbf{0}$. If the model is true, then $\mathbb{E}\mathbf{Y}$ is $\alpha\mathbf{1}$.

Figure 4.1 demonstrates the observations picture and the variables picture in this context. It includes everything from Fig. 2.1 along with additional items corresponding to the location model's assumption.

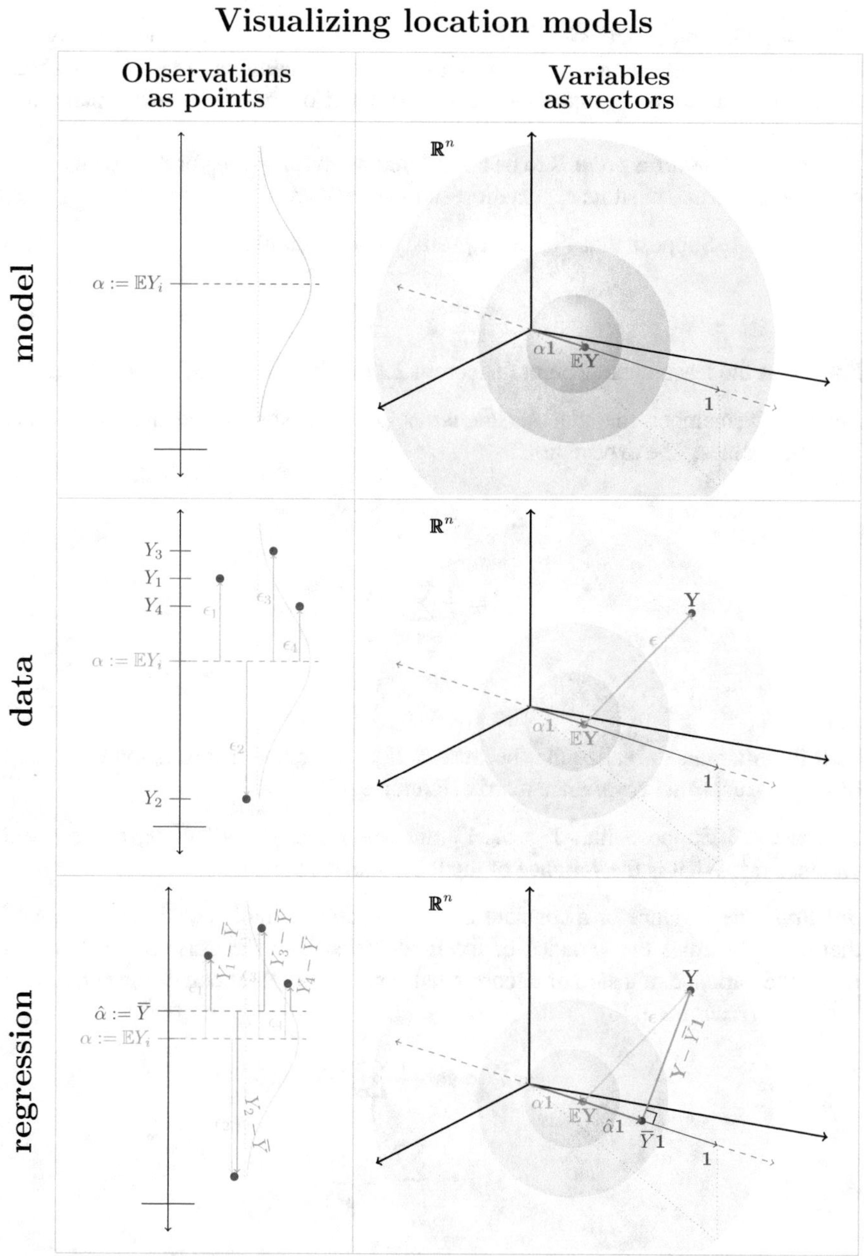


**Fig. 4.1** **Observations as points.** *model*: The response values share the same expectation $\alpha$. *data*: Arrows represent the random errors which "kick" the response values away from their expectation. *regression*: The least-squares point (average of the responses) can be considered an estimate of their expectation. **Variables as vectors.** *model*: A three-dimensional subspace is depicted that includes both the vectors $\mathbf{y}$ and $\mathbf{1}$. If the responses share the same expectation, then the expectation vector is in span$\{\mathbf{1}\}$. A density for the response is also depicted. *data*: The error random vector "kicks" the response vector away from its expectation. *regression*: The response is projected back into span$\{\mathbf{1}\}$ to arrive at the least-squares prediction vector $\hat{\alpha}\mathbf{1}$. The sample average $\hat{\alpha} = \overline{Y}$ can be considered an estimate of the common expectation $\alpha$

Notice that the *distribution* of $\mathbf{Y} = (Y_1, \ldots, Y_n)$ is not fully specified by the location model definition; there is room to bring in additional assumptions about the errors, for example, that they are uncorrelated or that they have a particular distribution.

If the *least-squares point* is to be considered as an *estimator* for the true location $\alpha$, we ought to ask what its expectation and variance are.

**Exercise 4.1** Suppose that $Y_1, \ldots, Y_n$ satisfy a location model

$$Y_i = \alpha + \epsilon_i .$$

Show that the least-squares point (Theorem 2.1) is an unbiased estimator for $\alpha$.

*Solution* Remember that the *least-squares point* is simply the average of the response values. The expectation is

$$\begin{aligned}\mathbb{E}\overline{Y} &= \mathbb{E}\left(\tfrac{1}{n}\sum_i Y_i\right) \\ &= \tfrac{1}{n}\sum_i \underbrace{\mathbb{E}Y_i}_{\alpha} \\ &= \alpha .\end{aligned}$$

♦

It is important to realize that because $\mathbf{Y}$ is a non-random translation of $\boldsymbol{\epsilon}$, they have the exact same covariance matrix (Exercise 3.23).

**Exercise 4.2** Suppose that $Y_1, \ldots, Y_n$ are uncorrelated and all have the same variance $\sigma^2$. What is the variance of the least-squares point?

*Solution* The variance of a constant times a random variable equals the square of that constant times the variance of the random variable (Exercise 3.23). Furthermore, the variance of a sum of uncorrelated random variables equals the sum of the variances (Exercise 3.26).

$$\begin{aligned}\operatorname{var}\overline{Y} &= \operatorname{var}\left(\tfrac{1}{n}\sum_i Y_i\right) \\ &= \tfrac{1}{n^2}\sum_i \underbrace{\operatorname{var}Y_i}_{\sigma^2} \\ &= \tfrac{1}{n^2}(n\sigma^2) \\ &= \frac{\sigma^2}{n} .\end{aligned}$$

♦

### 4.1.2 Simple Linear Models

**Definition (Simple Linear Model)** Let $Y_1, \ldots, Y_n$ be random variables representing a *response* variable and $x_1, \ldots, x_n \in \mathbb{R}$ be observations of an *explanatory* variable. A **simple linear model** assumes that each observation can be represented by

$$Y_i = \alpha + \beta(x_i - \bar{x}) + \epsilon_i,$$

where $\alpha, \beta \in \mathbb{R}$ and $\epsilon_1, \ldots, \epsilon_n$ are random variables that each have expectation 0.

In terms of vectors, the simple linear model assumption can be expressed with $\mathbf{Y} = (Y_1, \ldots, Y_n)$ as

$$\begin{aligned}\begin{bmatrix} Y_1 \\ \vdots \\ Y_n \end{bmatrix} &= \begin{bmatrix} \alpha + \beta(x_1 - \bar{x}) + \epsilon_1 \\ \vdots \\ \alpha + \beta(x_n - \bar{x}) + \epsilon_n \end{bmatrix} \\ &= \alpha \begin{bmatrix} 1 \\ \vdots \\ 1 \end{bmatrix} + \beta \left( \begin{bmatrix} x_1 \\ \vdots \\ x_n \end{bmatrix} - \bar{x} \begin{bmatrix} 1 \\ \vdots \\ 1 \end{bmatrix} \right) + \begin{bmatrix} \epsilon_1 \\ \vdots \\ \epsilon_n \end{bmatrix} \\ &= \alpha \mathbf{1} + \beta(\mathbf{x} - \bar{x}\mathbf{1}) + \boldsymbol{\epsilon},\end{aligned}$$

where $\boldsymbol{\epsilon} := (\epsilon_1, \ldots, \epsilon_n)$ is a random vector with expectation $\mathbf{0}$. If the model is true, then $\mathbb{E}\mathbf{Y}$ is $\alpha\mathbf{1} + \beta(\mathbf{x} - \bar{x}\mathbf{1})$.

Figure 4.2 shows the familiar scatterplot picture along with depictions of the response and explanatory vectors in $\mathbb{R}^n$. It includes everything from Fig. 2.2 along with additional items corresponding to the simple linear model's assumption.

**Exercise 4.3** Let $x_1, \ldots, x_n \in \mathbb{R}$ be values of an explanatory variable, and suppose that the response variable $Y_1, \ldots, Y_n$ satisfies a simple linear model

$$Y_i = \alpha + \beta(x_i - \bar{x}) + \epsilon_i.$$

Assuming $x_1, \ldots, x_n$ are all the same number, show that the coefficients $\hat{\alpha}$ and $\hat{\beta}$ in the least-squares line $y = \hat{\alpha} + \hat{\beta}(x - \bar{x})$ are unbiased estimators for $\alpha$ and $\beta$.

**Exercise 4.4** Suppose $Y_1, \ldots, Y_n$ are uncorrelated and all have the same variance $\sigma^2$. If $x_1, \ldots, x_n$ are values of an explanatory variable, what is the covariance matrix of the coefficients $\hat{\alpha}$ and $\hat{\beta}$ in the least-squares line $y = \hat{\alpha} + \hat{\beta}(x - \bar{x})$?

**Exercise 4.5** Suppose that a response variable satisfies a simple linear model of an explanatory variable and that it is predicted by the least-squares line. Which is

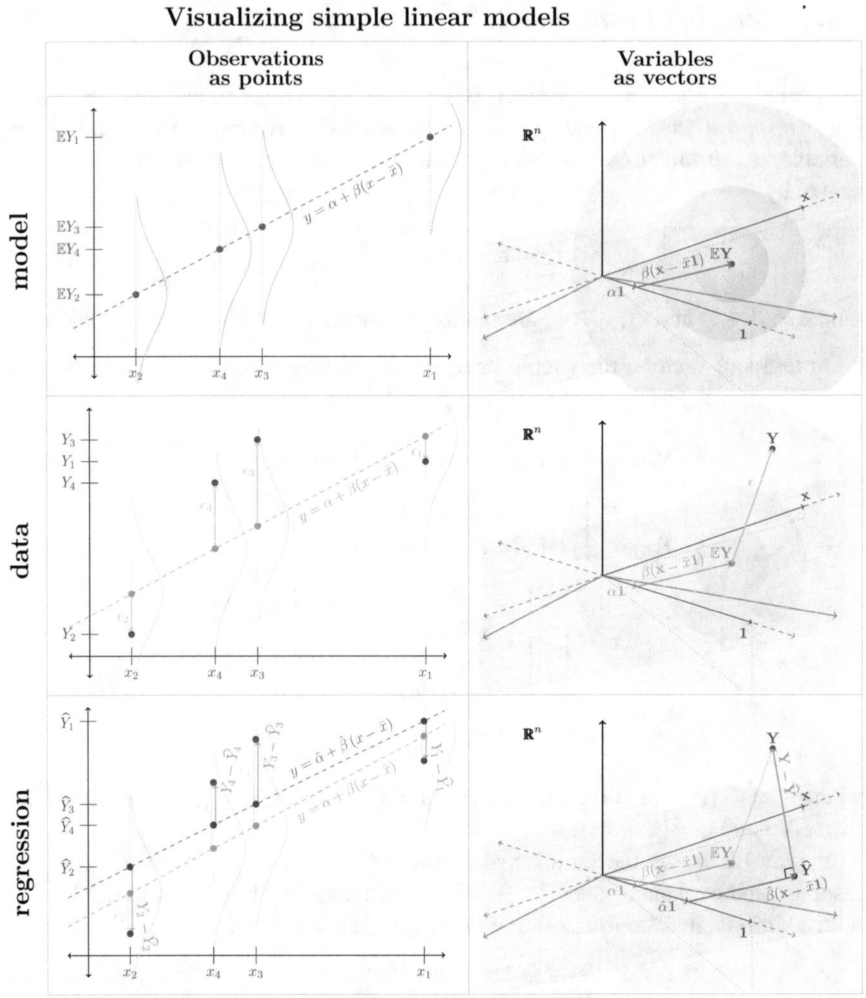


**Fig. 4.2** **Observations as points.** *model*: The response values have expectations along the line $\alpha + \beta(x - \bar{x})$. *data*: Arrows represent the random errors which "kick" the response values away from their expectations. *regression*: The least-squares line provides estimates of their expectations. **Variables as vectors.** *model*: A three-dimensional subspace is depicted that includes the vectors $\mathbf{y}$, $\mathbf{1}$, and $\mathbf{x}$. If the responses' expectations are along a line, then the expectation vector is in span$\{\mathbf{1}, \mathbf{x}\}$. A density for the response is also depicted. *data*: The error random vector "kicks" the response vector away from its expectation. *regression*: The response is projected back into span$\{\mathbf{1}, \mathbf{x}\}$ to arrive at the least-squares prediction vector $\hat{\alpha}\mathbf{1} + \hat{\beta}(\mathbf{x} - \bar{x}\mathbf{1})$. The coefficients $\hat{\alpha}$ and $\hat{\beta}$ can be considered estimates of the coefficients $\alpha$ and $\beta$ from the true expectations' line

larger: the sum of squared *errors* or the sum of squared *residuals*? Base your answer on the definition of the least-squares line, and explain.

*Solution* The sum of squared errors is the sum of squared differences between the response values and the true line, while the sum of squared residuals is the sum of squared differences between the points and the least-squares line. The least-squares line is, by definition, the one with the smallest possible sum of squared differences from the points, so the sum of squared residuals cannot possibly be larger than the sum of squared errors. ♦

**Exercise 4.6** The variables picture provides us with a more specific answer to the question posed in Exercise 4.5. Use the Pythagorean identity to quantify the difference between the sum of squared errors and the sum of squared residuals.

*Solution* The error vector forms the hypotenuse of a right triangle whose other sides are $\widehat{\mathbf{Y}} - \mathbb{E}\mathbf{Y}$ and the residual vector $\mathbf{Y} - \widehat{\mathbf{Y}}$. Invoking the Pythagorean identity,

$$\|\boldsymbol{\epsilon}\|^2 = \|\mathbf{Y} - \widehat{\mathbf{Y}}\|^2 + \|\widehat{\mathbf{Y}} - \mathbb{E}\mathbf{Y}\|^2.$$

The sum of squared errors is larger than the sum of squared residuals by $\|\widehat{\mathbf{Y}} - \mathbb{E}\mathbf{Y}\|^2$. ♦

### 4.1.3 Multiple Linear Models

**Definition (Multiple Linear Model)** Let $Y_1, \ldots, Y_n$ be random variables representing a *response* variable and $(x_1^{(1)}, \ldots, x_1^{(m)}), \ldots, (x_n^{(1)}, \ldots, x_n^{(m)}) \in \mathbb{R}^m$ be $n$ observations of $m$ *explanatory* variables. A **multiple linear model** assumes that each observation can be represented by

$$Y_i = \alpha + \beta_1(x_i^{(1)} - \bar{x}^{(1)}) + \ldots + \beta_m(x_i^{(m)} - \bar{x}^{(m)}) + \epsilon_i,$$

where $\alpha, \beta_1, \ldots, \beta_m \in \mathbb{R}$ and $\epsilon_1, \ldots, \epsilon_n$ are random variables that each have expectation 0.

In terms of vectors, the multiple linear model assumption says that $\mathbf{Y} = (Y_1, \ldots, Y_n)$ satisfies

$$\begin{aligned}
\begin{bmatrix} Y_1 \\ \vdots \\ Y_n \end{bmatrix} &= \begin{bmatrix} \alpha + \beta_1(x_1^{(1)} - \bar{x}^{(1)}) + \ldots + \beta_m(x_1^{(m)} - \bar{x}^{(m)}) + \epsilon_1 \\ \vdots \\ \alpha + \beta_1(x_n^{(1)} - \bar{x}^{(1)}) + \ldots + \beta_m(x_n^{(m)} - \bar{x}^{(m)}) + \epsilon_n \end{bmatrix} \\
&= \alpha \begin{bmatrix} 1 \\ \vdots \\ 1 \end{bmatrix} + \beta_1 \begin{bmatrix} x_1^{(1)} - \bar{x}^{(1)} \\ \vdots \\ x_n^{(1)} - \bar{x}^{(1)} \end{bmatrix} + \ldots + \beta_m \begin{bmatrix} x_1^{(m)} - \bar{x}^{(m)} \\ \vdots \\ x_n^{(m)} - \bar{x}^{(m)} \end{bmatrix} + \begin{bmatrix} \epsilon_1 \\ \vdots \\ \epsilon_n \end{bmatrix} \\
&= \alpha \mathbf{1} + \widetilde{\mathbb{X}}\boldsymbol{\beta} + \boldsymbol{\epsilon},
\end{aligned}$$

where $\boldsymbol{\beta} = (\beta_1, \ldots, \beta_m) \in \mathbb{R}^m$, $\widetilde{\mathbb{X}}$ is the centered explanatory data matrix, and $\boldsymbol{\epsilon} := (\epsilon_1, \ldots, \epsilon_n)$ is a random vector with expectation $\mathbf{0}$. If the model is true, then $\mathbb{E}\mathbf{Y}$ is $\alpha\mathbf{1} + \widetilde{\mathbb{X}}\boldsymbol{\beta}$.

It takes a bit more care to draw the new version of our previous multiple linear regression picture (Fig. 2.3) in $\mathbb{R}^n$ that includes $\mathbb{E}\mathbf{Y}$, $\mathbf{Y}$, and $\widehat{\mathbf{Y}}$. The column space of $\mathbb{X}$ will be depicted as a generic two-dimensional subspace in the picture. In particular, the intersection of the subspace with our picture has to be the plane that includes the origin, $\mathbb{E}\mathbf{Y}$ and $\widehat{\mathbf{Y}}$. As a result, we are not at liberty to draw any specific column vector of $\mathbb{X}$ if we want the picture to remain accurate; see Fig. 4.3.

**Exercise 4.7** Let $(x_1^{(1)}, \ldots, x_1^{(m)}), \ldots, (x_n^{(1)}, \ldots, x_n^{(m)}) \in \mathbb{R}^m$ be $n$ observations of $m$ explanatory variables, and suppose that the response variable $Y_1, \ldots, Y_n$ satisfies a multiple linear model

$$Y_i = \alpha + \beta_1(x^{(1)} - \bar{x}^{(1)}) + \ldots + \beta_m(x^{(m)} - \bar{x}^{(m)}) + \epsilon_i.$$

Assuming the explanatory variables' empirical covariance matrix $\Sigma$ is full rank, show that the coefficients $\hat{\alpha}, \hat{\beta}_1, \ldots, \hat{\beta}_m$ in the least-squares hyperplane $y = \hat{\alpha} + \hat{\beta}_1(x^{(1)} - \bar{x}^{(1)}) + \ldots + \hat{\beta}_m(x^{(m)} - \bar{x}^{(m)})$ are unbiased estimators for $\alpha, \beta_1, \ldots, \beta_m$.

*Solution* The *least-squares hyperplane* has $\hat{\alpha} = \overline{Y}$, which can be expressed as

$$\begin{aligned}
\overline{Y} &= \tfrac{1}{n}\sum_i Y_i \\
&= \tfrac{1}{n}\sum_i [\alpha + \beta_1(x_i^{(1)} - \bar{x}^{(1)}) + \ldots + \beta_m(x_i^{(m)} - \bar{x}^{(m)}) + \epsilon_i] \\
&= \alpha + \beta_1 \underbrace{\tfrac{1}{n}\sum_i (x_i^{(1)} - \bar{x}^{(1)})}_{0} + \ldots + \beta_m \underbrace{\tfrac{1}{n}\sum_i (x_i^{(m)} - \bar{x}^{(m)})}_{0} + \tfrac{1}{n}\sum_i \epsilon_i \\
&= \alpha + \tfrac{1}{n}\sum_i \epsilon_i.
\end{aligned}$$

Its expectation is

$$\begin{aligned}
\mathbb{E}\overline{Y} &= \alpha + \tfrac{1}{n}\sum_i \underbrace{\mathbb{E}\epsilon_i}_{0} \\
&= \alpha.
\end{aligned}$$

The vector of empirical covariances of $\mathbf{Y}$ with $\mathbf{x}^{(1)}, \ldots, \mathbf{x}^{(m)}$ can be expressed as $\frac{1}{n}\widetilde{\mathbb{X}}^T\mathbf{Y}$ where $\widetilde{\mathbb{X}}$ is the centered version of the explanatory data matrix. Substituting this representation into the formula from Theorem 2.3,

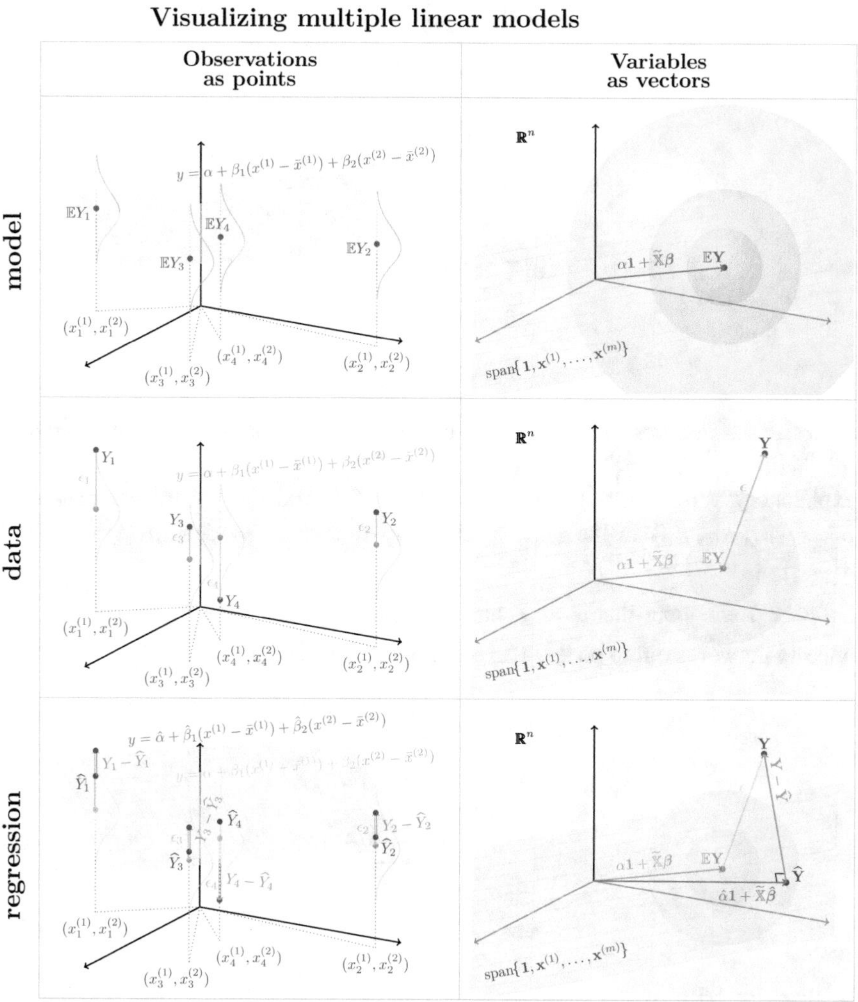


**Fig. 4.3** **Observations as points.** *model*: The response values have expectations along a plane. *data*: Arrows represent the random errors which "kick" the response values away from their expectations. *regression*: The least-squares plane has the smallest possible sum of squared residuals. **Variables as vectors.** *model*: A three-dimensional subspace is depicted that includes $\mathbf{y}$ and a two-dimensional subspace of $\text{span}\{\mathbf{1}, \mathbf{x}^{(1)}, \dots, \mathbf{x}^{(m)}\}$. If the responses' expectations are along a hyperplane, then the expectation vector is in $\text{span}\{\mathbf{1}, \mathbf{x}^{(1)}, \dots, \mathbf{x}^{(m)}\}$. A density for the response is also depicted. *data*: The error random vector "kicks" the response vector away from its expectation. *regression*: The response is projected back into $\text{span}\{\mathbf{1}, \mathbf{x}^{(1)}, \dots, \mathbf{x}^{(m)}\}$ to arrive at the least-squares prediction vector. Its coefficients can be considered estimates of the corresponding coefficients from the true expectations' hyperplane

$$\begin{aligned}
\mathbb{E}\hat{\boldsymbol{\beta}} &= \mathbb{E}\Sigma^{-1}\tfrac{1}{n}\widetilde{\mathbb{X}}^T\mathbf{Y} \\
&= \Sigma^{-1}\tfrac{1}{n}\widetilde{\mathbb{X}}^T\mathbb{E}\mathbf{Y} \\
&= \Sigma^{-1}\tfrac{1}{n}\widetilde{\mathbb{X}}^T(\alpha\mathbf{1}+\widetilde{\mathbb{X}}\boldsymbol{\beta}) \\
&= \Sigma^{-1}(\tfrac{\alpha}{n}\underbrace{\widetilde{\mathbb{X}}^T\mathbf{1}}_{0}+\underbrace{\tfrac{1}{n}\widetilde{\mathbb{X}}^T\widetilde{\mathbb{X}}}_{\Sigma}\boldsymbol{\beta}) \\
&= \Sigma^{-1}\Sigma\boldsymbol{\beta} \\
&= \boldsymbol{\beta}.
\end{aligned}$$

♦

**Exercise 4.8** Suppose $Y_1, \ldots, Y_n$ are uncorrelated and all have the same variance $\sigma^2$. With $(x_1^{(1)}, \ldots, x_1^{(m)}), \ldots, (x_n^{(1)}, \ldots, x_n^{(m)}) \in \mathbb{R}^m$ as $n$ observations of $m$ explanatory variables, what is the variance of $\hat{\alpha}$ and the covariance matrix of $\hat{\boldsymbol{\beta}} = (\hat{\beta}_1, \ldots, \hat{\beta}_m)$ in the least-squares hyperplane[2] $y = \hat{\alpha} + \hat{\beta}_1(x^{(1)} - \bar{x}^{(1)}) + \ldots + \hat{\beta}_m(x^{(m)} - \bar{x}^{(m)})$?

*Solution* Remember that $\hat{\alpha} = \overline{Y}$ has the representation $\alpha + \frac{1}{n}\sum_i \epsilon_i$. Its variance once again works out to be $\frac{\sigma^2}{n}$. The covariance matrix of $\hat{\boldsymbol{\beta}}$ is

$$\begin{aligned}
\operatorname{cov}\hat{\boldsymbol{\beta}} &= \operatorname{cov}\Sigma^{-}\tfrac{1}{n}\widetilde{\mathbb{X}}^T\mathbf{Y} \\
&= \Sigma^{-}\tfrac{1}{n}\widetilde{\mathbb{X}}^T(\sigma^2\mathbb{I})[\Sigma^{-}\tfrac{1}{n}\widetilde{\mathbb{X}}^T]^T \\
&= \tfrac{\sigma^2}{n}\Sigma^{-}\underbrace{(\tfrac{1}{n}\widetilde{\mathbb{X}}^T\widetilde{\mathbb{X}})}_{\Sigma}\Sigma^{-} \\
&= \tfrac{\sigma^2}{n}\Sigma^{-}
\end{aligned}$$

by Exercise 1.75. ♦

**Exercise 4.9** Suppose $Y_1, \ldots, Y_n$ are uncorrelated and all have the same variance $\sigma^2$. With $(x_1^{(1)}, \ldots, x_1^{(m)}), \ldots, (x_n^{(1)}, \ldots, x_n^{(m)}) \in \mathbb{R}^m$ as $n$ observations of $m$ explanatory variables, show that $\hat{\alpha}$ is uncorrelated with every $\hat{\beta}_1, \ldots, \hat{\beta}_m$ in the least-squares hyperplane.

Let us think about how to interpret the formula for the least-squares coefficients' covariance matrix that we worked out in Exercise 4.8. The $\sigma^2$ in the formula tells us

[2]Here and throughout the text, we may speak of "the" least-squares coefficients without assuming that there is a unique coefficient vector that minimizes the sum of squared residuals. In such cases, the least-squares coefficients will refer to those obtained using the Moore–Penrose inverse as in Theorems 2.3 and 2.4.

that the variance in estimating the coefficients is proportional to the error variance; this should seem intuitive. It is also intuitive that the variance in estimating the coefficients should decrease with sample size. In fact, if you double the sample size by bringing in new observations with all the same explanatory values as the original data but with new response values, the variance in estimating the coefficients is cut in half. (By copying all the same explanatory values, the empirical covariance matrix is unchanged.) Finally, one can interpret the role of the explanatory data's empirical covariance matrix in light of our discussion of principal components in Sect. 1.15 or the generalization of that discussion in Sects. 3.4 and 3.5. The *directions* of high empirical variance of the explanatory data have low variance in the corresponding least-squares estimator, and vice versa.

### 4.1.4 Linear Models

**Definition (Linear Model)** Let $Y_1, \dots, Y_n$ be random variables representing a *response* variable and $\mathbf{x}_1, \dots, \mathbf{x}_n \in \mathbb{R}^m$ be observations of $m$ *explanatory* variables. A **linear model** assumes that each observation can be represented by

$$Y_i = \gamma_1 g_1(\mathbf{x}_i) + \dots + \gamma_d g_d(\mathbf{x}_i) + \epsilon_i,$$

where $\gamma_1, \dots, \gamma_d \in \mathbb{R}$ and $\epsilon_1, \dots, \epsilon_n$ are random variables that each have expectation 0.

In terms of vectors, the linear model assumption says that $\mathbf{Y} = (Y_1, \dots, Y_n)$ satisfies

$$\begin{aligned}\begin{bmatrix} Y_1 \\ \vdots \\ Y_n \end{bmatrix} &= \begin{bmatrix} \gamma_1 g_1(\mathbf{x}_1) + \dots + \gamma_d g_d(\mathbf{x}_1) + \epsilon_1 \\ \vdots \\ \gamma_1 g_1(\mathbf{x}_n) + \dots + \gamma_d g_d(\mathbf{x}_n) + \epsilon_n \end{bmatrix} \\ &= \mathbb{M}\boldsymbol{\gamma} + \boldsymbol{\epsilon},\end{aligned}$$

where $\boldsymbol{\epsilon} := (\epsilon_1, \dots, \epsilon_n)$ is a random vector with expectation $\mathbf{0}$. If the model is true, then the expectation of $\mathbf{Y}$ is $\mathbb{M}\boldsymbol{\beta}$. We will call this $\mathbb{M}$ a *model design matrix* as opposed to a *regression design matrix* when it is necessary to distinguish them.

As with *linear* regression, *linear* modeling does not refer to the response and explanatory variables being related by a line. A model is linear if the expectation of the response is linear *in the free parameters*. So for instance, with $\theta$ ranging over $\mathbb{R}$, the model $Y_i = \theta e^{x_i} + \epsilon_i$ is a linear model, but $Y_i = e^{\theta x_i} + \epsilon_i$ is not.

Speaking of *linearity*, think about how convenient it is that the least-squares coefficient vector and the prediction vector are linear functions of the response vector. Not only does that make them easy to calculate, but it also makes them relatively easy to analyze, enabling us to readily derive expectations, covariances, and more throughout this chapter.

**Exercise 4.10** Let $\mathbf{x}_1, \ldots, \mathbf{x}_n \in \mathbb{R}^m$ be $n$ observations of $m$ explanatory variables, and suppose that the response variable $Y_1, \ldots, Y_n$ satisfies a linear model

$$Y_i = \gamma_1 g_1(\mathbf{x}_i) + \ldots + \gamma_d g_d(\mathbf{x}_i) + \epsilon_i.$$

Assuming the columns of the design matrix are linearly independent, show that the coefficients $\hat{\gamma}_1, \ldots, \hat{\gamma}_d$ in the least-squares linear fit $y = \hat{\gamma}_1 g_1(\mathbf{x}) + \ldots + \hat{\gamma}_d g_d(\mathbf{x})$ are unbiased estimators for $\gamma_1, \ldots, \gamma_d$.

*Solution* Let $\mathbb{M}$ represent the design matrix

$$\mathbb{M} := \begin{bmatrix} g_1(\mathbf{x}_1) & \cdots & g_d(\mathbf{x}_1) \\ \vdots & \ddots & \vdots \\ g_1(\mathbf{x}_n) & \cdots & g_d(\mathbf{x}_n) \end{bmatrix}.$$

The expectation of $\mathbf{Y} = \mathbb{M}\boldsymbol{\gamma} + \boldsymbol{\epsilon}$ is $\mathbb{M}\boldsymbol{\gamma}$. Using the formula for the least-squares coefficients provided in Theorem 2.4,

$$\begin{aligned} \mathbb{E}\hat{\boldsymbol{\gamma}} &= \mathbb{E}(\mathbb{M}^T\mathbb{M})^{-1}\mathbb{M}^T\mathbf{Y} \\ &= (\mathbb{M}^T\mathbb{M})^{-1}\mathbb{M}^T \underbrace{\mathbb{E}\mathbf{Y}}_{\mathbb{M}\boldsymbol{\gamma}} \\ &= (\mathbb{M}^T\mathbb{M})^{-1}(\mathbb{M}^T\mathbb{M})\boldsymbol{\gamma} \\ &= \boldsymbol{\gamma}. \end{aligned}$$

(We know that $\mathbb{M}^T\mathbb{M}$ is invertible because the columns of $\mathbb{M}$ are assumed to be linearly independent—see Exercise 1.63.) ♦

Keep in mind that the *regression* design matrix used for least-squares linear regression may not be the same as the *model* design matrix that actually defines the expectation of the response vector. Exercise 4.10 shows us that *when they coincide*, the least-squares coefficients are unbiased for the true coefficients. On the other hand, the *covariance matrix* of the least-squares coefficients does not depend on the response vector's expectation at all.

**Exercise 4.11** Suppose $Y_1, \ldots, Y_n$ are uncorrelated and all have the same variance $\sigma^2$. With $\mathbf{x}_1, \ldots, \mathbf{x}_n \in \mathbb{R}^m$ as $n$ observations of $m$ explanatory variables, what is the covariance matrix of $\hat{\boldsymbol{\gamma}} = (\hat{\gamma}_1, \ldots, \hat{\gamma}_d)$ in the least-squares linear fit $y = \hat{\gamma}_1 g_1(\mathbf{x}) + \ldots + \hat{\gamma}_d g_d(\mathbf{x})$?

*Solution* The covariance matrix of $\hat{\boldsymbol{\gamma}}$ is

$$\begin{aligned}\operatorname{cov}\hat{\boldsymbol{\gamma}} &= \operatorname{cov}(\mathbb{M}^T\mathbb{M})^-\mathbb{M}^T\mathbf{Y}\\ &= (\mathbb{M}^T\mathbb{M})^-\mathbb{M}^T(\sigma^2\mathbb{I})[(\mathbb{M}^T\mathbb{M})^-\mathbb{M}^T]^T\\ &= \sigma^2(\mathbb{M}^T\mathbb{M})^-\mathbb{M}^T\mathbb{M}(\mathbb{M}^T\mathbb{M})^-\\ &= \sigma^2(\mathbb{M}^T\mathbb{M})^-\end{aligned}$$

by Exercise 1.75. ♦

To clarify the role of sample size in the quality of estimating the linear model coefficients, one can also consider the expression $\frac{\sigma^2}{n}(\frac{1}{n}\mathbb{M}^T\mathbb{M})^-$ which is equivalent to the answer of Exercise 4.11. $\frac{1}{n}\mathbb{M}^T\mathbb{M}$ is a *second moments matrix* (Exercise 1.82), and it remains the same if an explanatory dataset is replicated. Thus with $k$ copies of a particular explanatory dataset (and new draws of the response variable for each copy), the variances and covariances of the least-squares coefficients will be $1/k$ times as large as they would have been with just one copy.

For an unbiased estimator, the smaller the variance the better. According to the following much-celebrated result, no linear (in the response vector) unbiased estimator for the coefficients is better than the least-squares estimator.

**Theorem 4.1 (Gauss–Markov theorem, Christensen (2011) Thm 2.3.2)** *Assume the columns of $\mathbb{M}$ are linearly independent. Suppose $\mathbf{Y} = \mathbb{M}\boldsymbol{\gamma} + \boldsymbol{\epsilon}$ where $\boldsymbol{\epsilon}$ is a random vector with mean $\mathbf{0}$ and covariance $\sigma^2\mathbb{I}$. Let $\hat{\boldsymbol{\gamma}} := (\mathbb{M}^T\mathbb{M})^{-1}\mathbb{M}^T\mathbf{Y}$ denote the least-squares estimator for the coefficients. Suppose an alternative estimator $\check{\boldsymbol{\gamma}} := \mathbb{M}\mathbf{Y}$ is also unbiased for $\boldsymbol{\gamma}$. Then the variance of $\check{\boldsymbol{\gamma}}^T\mathbf{v}$ is at least as large as the variance of $\hat{\boldsymbol{\gamma}}^T\mathbf{v}$ for every $\mathbf{v}$.*

**Exercise 4.12** Assume the columns of $\mathbb{M}$ are linearly independent. Suppose $\mathbf{Y} = \mathbb{M}\boldsymbol{\gamma} + \boldsymbol{\epsilon}$ where $\boldsymbol{\epsilon}$ is a random vector with mean $\mathbf{0}$ and covariance $\sigma^2\mathbb{I}$. Let $\hat{\boldsymbol{\gamma}} := (\mathbb{M}^T\mathbb{M})^{-1}\mathbb{M}^T\mathbf{Y}$ denote the least-squares estimator for the coefficients. Suppose an alternative estimator $\check{\boldsymbol{\gamma}} := \mathbb{L}\mathbf{Y}$ is also unbiased for $\boldsymbol{\gamma}$. Use the Gauss–Markov theorem to show that $\mathbb{E}(\hat{\boldsymbol{\gamma}}-\boldsymbol{\gamma})^T\mathbb{L}(\hat{\boldsymbol{\gamma}}-\boldsymbol{\gamma}) \le \mathbb{E}(\check{\boldsymbol{\gamma}}-\boldsymbol{\gamma})^T\mathbb{L}(\check{\boldsymbol{\gamma}}-\boldsymbol{\gamma})$ for every positive semi-definite matrix $\mathbb{L}$.

#### 4.1.4.1 Modeling with Categorical Explanatory Variables

Suppose there is only one explanatory variable $z_1, \ldots, z_n$ and that it is categorical, taking values in $\{1, \ldots, k\}$. Let $\alpha_j$ denote the expectation of draws of the response variable from group $j$. This scenario can be written as a linear model:

$$Y_i = \alpha_1 \mathbb{1}_{z_i=1} + \ldots + \alpha_k \mathbb{1}_{z_i=k} + \epsilon_i.$$

Remember that least-squares predicts each group by the average of the response values from that group, as described at the end of Chap. 2. We will come back to this scenario for hypothesis testing in Sect. 6.2.4.1.

And as in Chap. 2, linear models can combine categorical and quantitative explanatory variables. For example, if $\mathbf{x}$ is a quantitative explanatory variable, $\mathbf{z}$ is a categorical explanatory variable, and $\mathbf{Y}$ is a quantitative response variable, you could assume that each group has its own separate line relating $x$-values to expectation of response values. Written in terms of all the variables together, this is called an *interactions* model.

$$Y_i = \alpha_1 \mathbb{1}_{z_i=1} + \ldots + \alpha_k \mathbb{1}_{z_i=k} + \beta_1(x_i - \bar{x})\mathbb{1}_{z_i=1} + \ldots + \beta_k(x_i - \bar{x})\mathbb{1}_{z_i=k} + \epsilon_i.$$

Alternatively, a *simpler* model of the data assumes that all groups' lines share the same slope. That case is called an *additive* model.

$$Y_i = \alpha_1 \mathbb{1}_{z_i=1} + \ldots + \alpha_k \mathbb{1}_{z_i=k} + \beta(x_i - \bar{x}) + \epsilon_i.$$

#### 4.1.4.2 Predicting a New Response Value

It may not be the coefficients themselves that are important. Rather, the purpose of linear modeling is often to *predict* future response values using future explanatory values. Let us consider a new observation generated according to the same mechanism as the original data:

$$Y_{n+1} = \gamma_1 g_1(\mathbf{x}_{n+1}) + \ldots + \gamma_d g_d(\mathbf{x}_{n+1}) + \epsilon_{n+1},$$

where $\epsilon_{n+1}$ is uncorrelated with the previous errors $\epsilon_1, \ldots, \epsilon_n$ and has the same variance $\sigma^2$.

An obvious choice for predicting $Y_{n+1}$ is to substitute the least-squares linear regression coefficients

$$\widehat{Y}_{n+1} := \hat{\gamma}_1 g_1(\mathbf{x}_{n+1}) + \ldots + \hat{\gamma}_d g_d(\mathbf{x}_{n+1}).$$

It is easy to see that this random variable is an unbiased estimator for $\mathbb{E}\widehat{Y}_{n+1}$ because the least-squares coefficients are unbiased for the true coefficients.

$$\begin{aligned}
\mathbb{E}\widehat{Y}_{n+1} &= \mathbb{E}[\hat{\gamma}_1 g_1(\mathbf{x}_{n+1}) + \ldots + \hat{\gamma}_d g_d(\mathbf{x}_{n+1})] \\
&= \gamma_1 g_1(\mathbf{x}_{n+1}) + \ldots + \gamma_d g_d(\mathbf{x}_{n+1}) \\
&= \mathbb{E}Y_{n+1}
\end{aligned}$$

as long as the original explanatory data had full rank covariance. Actually, full rank covariance is not necessary for the prediction to have the same expectation as the new response value, as Exercise 4.13 shows. Section 4.2.2 further explores the quality of this prediction.

**Exercise 4.13** Suppose $\mathbf{Y} = \mathbb{M}\gamma + \epsilon$ with $\mathbb{E}\epsilon = 0$, and let $\hat{\gamma}$ be coefficients of least-squares linear regression estimators for the correctly specified model. If a new explanatory observation $\mathbf{v}_{n+1}$ is in the row space of $\mathbb{M}$, and $Y_{n+1} = \mathbf{v}_{n+1}^T\gamma + \epsilon_{n+1}$ with $\mathbb{E}\epsilon_{n+1} = 0$, show that the expectation of the predictor $\widehat{Y}_{n+1} = \mathbf{v}_{n+1}^T\hat{\gamma}$ equals the expectation of $Y_{n+1}$ regardless of whether or not $\mathbb{M}$ has full rank.

#### 4.1.4.3 Identifiability

In Chap. 2, we noted a potential pathology of the explanatory data. If every component of the $\mathbf{x}$ vector is the same, then the span of $\{\mathbf{1}, \mathbf{x}\}$ is one-dimensional rather than two-dimensional. In that case, there are infinitely many linear combinations of $\mathbf{1}$ and $\mathbf{x} - \bar{x}\mathbf{1}$ that result in the orthogonal projection $\overline{Y}\mathbf{1}$. Likewise there are infinitely many coefficient pairs $(a, b)$ for which $\mathbb{E}\mathbf{Y} = a\mathbf{1} + b(\mathbf{x} - \bar{x}\mathbf{1})$. More generally, if the columns of a regression design matrix are not linearly independent, then there are infinitely many coefficient vectors that minimize the sum of squared residuals; if the columns of a model design matrix are not linearly independent, then there are infinitely many coefficient vectors that imply the exact same expectation of $\mathbf{Y}$ via the linear model equation.

**Definition (Identifiable)** A model is **identifiable** if every value of the parameter space implies a different distribution for the response data.

If a model is not identifiable, that is, there are multiple possible values of the parameters that would say exactly the same thing about the data's distribution, then we could not possibly hope to tell which one is the *true* parameter value no matter how much data we have. However, even if the full parameter vector cannot be estimated, certain linear combinations of the parameters can be.

**Exercise 4.14** Let $\mathbf{x} = (x_1, \ldots, x_n)$ and $\mathbf{z} = (z_1, \ldots, z_n)$ be explanatory variables such that $\mathbf{x} = a\mathbf{z}$ for some $c \in \mathbb{R}$. Assume $\mathbf{Y} = (Y_1, \ldots, Y_n)$ satisfy $\mathbf{Y} = b_1\mathbf{x} + b_2\mathbf{z} + \epsilon$ for some $b_1, b_2 \in \mathbb{R}$ and $\mathbb{E}\epsilon = \mathbf{0}$. Argue that the derived parameter $c := ab_1 + b_2$ can be estimated.

## 4.2 Expected Sums of Squares

So far, we have introduced models that do not prescribe a specific *distribution* for the error vector. Still, we were able to derive the expectations and covariances matrices of random vectors that arise from doing least-squares linear regression. Interestingly, one can also analyze the expected squared norms of several important vectors without needing to know the distributions of the errors.

### 4.2.1 Estimating Error Variance

We have seen how to estimate the coefficients, but there is one remaining parameter that we need to discuss. Many of the exercises have assumed that the errors have covariance matrix $\sigma^2\mathbb{I}$ for some unspecified $\sigma^2$; this variance of the errors $\sigma^2$, called the *error variance*, is generally assumed to be unknown. We will now work out an estimator for the error variance that will play a crucial role in Chap. 6.

Let $\widehat{\mathbf{Y}}$ be the orthogonal projection of $\mathbf{Y}$ onto the column space of $\mathbb{M}$; let us think about its relationship to the error vector $\boldsymbol{\epsilon} = \mathbf{Y} - \mathbb{E}\mathbf{Y}$. We can express the error vector as

$$\begin{aligned}\boldsymbol{\epsilon} &= \mathbf{Y} - \mathbb{E}\mathbf{Y} \\ &= \underbrace{(\mathbf{Y} - \widehat{\mathbf{Y}})}_{\perp C(\mathbb{M})} - (\mathbb{E}\mathbf{Y} - \widehat{\mathbf{Y}}).\end{aligned}$$

When $\mathbb{E}\mathbf{Y}$ is in $C(\mathbb{M})$, so is $\mathbb{E}\mathbf{Y} - \widehat{\mathbf{Y}}$ and the above equation reveals two orthogonal projections of $\boldsymbol{\epsilon}$. It shows that the least-squares residual vector $\mathbf{Y} - \widehat{\mathbf{Y}}$ is the orthogonal projection of $\boldsymbol{\epsilon}$ onto $C(\mathbb{M})^{\perp}$. It also shows that $\widehat{\mathbf{Y}} - \mathbb{E}\mathbf{Y}$ is the orthogonal projection of $\boldsymbol{\epsilon}$ onto $C(\mathbb{M})$. The latter fact can also be expressed as $\widehat{\mathbf{Y}} = \mathbb{E}\mathbf{Y} + \mathbb{H}\boldsymbol{\epsilon}$ where $\mathbb{H}$ is the orthogonal projection matrix onto $C(\mathbb{M})$.

**Exercise 4.15** Suppose $\mathbf{Y} = \mathbb{M}\boldsymbol{\beta} + \boldsymbol{\epsilon}$ with $\operatorname{cov}\boldsymbol{\epsilon} = \sigma^2\mathbb{I}_n$, and let $\widehat{\mathbf{Y}}$ be the orthogonal projection of $\mathbf{Y}$ onto $C(\mathbb{M})$. Find $\mathbb{E}\|\mathbf{Y} - \widehat{\mathbf{Y}}\|^2$, the expected sum of squared residuals.

*Solution* We will let $\mathbb{H}$ be the orthogonal projection matrix onto $\mathbb{M}$, and use Exercise 3.35 along with Exercises 1.67 and 1.68.

$$\begin{aligned}\mathbb{E}\|\mathbf{Y} - \widehat{\mathbf{Y}}\|^2 &= \|(\mathbb{I} - \mathbb{H})\boldsymbol{\epsilon}\|^2 \\ &= \operatorname{tr}[(\mathbb{I} - \mathbb{H})\sigma^2\mathbb{I}] \\ &= \sigma^2(n - \operatorname{rank}\mathbb{M}).\end{aligned}$$

♦

The result of Exercise 4.15 provides us with an opportunity to estimate the error variance. Assuming the model is correctly specified (the regression design matrix is also the true model's design matrix) and the errors have covariance matrix $\sigma^2\mathbb{I}$, the estimator

$$\hat{\sigma}^2 := \frac{\|\mathbf{Y} - \widehat{\mathbf{Y}}\|^2}{n - \operatorname{rank}\mathbb{M}}$$

has expectation $\sigma^2$.

### 4.2.2 Prediction Error

Let us return to the context of predicting a new observation from a linear model as in Sect. 4.1.4.2. We saw that $\widehat{Y}_{n+1}$ has the same expectation as $Y_{n+1}$, but that does not really tell us that its predictions are close to the actual values. To get a sense of how good the prediction tends to be, we will analyze the *expected squared prediction error* $\mathbb{E}(Y_{n+1} - \widehat{Y}_{n+1})^2$. Because the prediction error has expectation zero, this expected squared difference is the variance of $Y_{n+1} - \widehat{Y}_{n+1}$. Observe that $\epsilon_{n+1}$ (and therefore $Y_{n+1}$) is uncorrelated with $\widehat{Y}_{n+1}$ which is within a constant of a linear combination of $\epsilon_1, \ldots, \epsilon_n$ (Exercise 3.12). With this in mind, we can neatly decompose the expected squared prediction error into the variance of $Y_{n+1}$ (which makes $Y_{n+1}$ inherently unpredictable) plus the variance of the predictor $\widehat{Y}_{n+1}$.

$$\begin{aligned}\mathbb{E}(Y_{n+1} - \widehat{Y}_{n+1})^2 &= \text{var}\,(Y_{n+1} - \widehat{Y}_{n+1})\\ &= \underbrace{\text{var}\ Y_{n+1}}_{\sigma^2} + \text{var}\ \widehat{Y}_{n+1}.\end{aligned}$$

This type of decomposition is not specific to linear modeling. If you replace the uncorrelated errors assumption with a stronger assumption that the observations are independent, you obtain a very general result (Exercise 4.16) which reveals that *predicting a new response as well as possible reduces to estimating its expectation as well as possible*.

**Exercise 4.16** Suppose the predictor $\widehat{Y}_{n+1}$ is a function of $Y_1, \ldots, Y_n$ which are *independent* of $Y_{n+1}$. Show that

$$\mathbb{E}(Y_{n+1} - \widehat{Y}_{n+1})^2 = \text{var}\ Y_{n+1} + \mathbb{E}(\widehat{Y}_{n+1} - \mathbb{E}Y_{n+1})^2.$$

Let us specialize to the multiple linear modeling context and derive a meaningful expression for the variance of $\widehat{Y}_{n+1}$. With the vector of new explanatory values denoted $\mathbf{x}_{n+1} = (x_{n+1}^{(1)}, \ldots, x_{n+1}^{(m)})$ and the empirical mean vector denoted $\bar{\mathbf{x}} = (\bar{x}^{(1)}, \ldots, \bar{x}^{(m)})$, the prediction can be expressed as

$$\begin{aligned}\widehat{Y}_{n+1} &= \hat{\alpha} + \hat{\beta}_1(x_{n+1}^{(1)} - \bar{x}^{(1)}) + \ldots + \hat{\beta}_m(x_{n+1}^{(m)} - \bar{x}^{(m)})\\ &= \hat{\alpha} + (\mathbf{x}_{n+1} - \bar{\mathbf{x}})^T\hat{\boldsymbol{\beta}}.\end{aligned}$$

Exercise 4.9 tells us that $\hat{\alpha}$ is uncorrelated with the other least-squares coefficients, so

$$\begin{aligned}\text{var}\ \widehat{Y}_{n+1} &= \text{var}\,[\hat{\alpha} + (\mathbf{x}_{n+1} - \bar{\mathbf{x}})^T\hat{\boldsymbol{\beta}}]\\ &= \underbrace{\text{var}\ \hat{\alpha}}_{\sigma^2/n} + \text{var}\,(\mathbf{x}_{n+1} - \bar{\mathbf{x}})^T\hat{\boldsymbol{\beta}}.\end{aligned}$$

The variance of a random variable can equivalently be considered the $1 \times 1$ covariance matrix of an $\mathbb{R}^1$-valued random vector. When we write the variance of the second term as a covariance, we can apply Exercise 3.23 to pull out a non-random $1 \times m$ matrix.

$$\begin{aligned}
\operatorname{var}(\mathbf{x}_{n+1} - \bar{\mathbf{x}})^T \hat{\boldsymbol{\beta}} &= \operatorname{cov}(\mathbf{x}_{n+1} - \bar{\mathbf{x}})^T \hat{\boldsymbol{\beta}} \\
&= (\mathbf{x}_{n+1} - \bar{\mathbf{x}})^T (\operatorname{cov} \hat{\boldsymbol{\beta}})(\mathbf{x}_{n+1} - \bar{\mathbf{x}}) \\
&= \frac{\sigma^2}{n} \underbrace{(\mathbf{x}_{n+1} - \bar{\mathbf{x}})^T \Sigma^{-1} (\mathbf{x}_{n+1} - \bar{\mathbf{x}})}_{\text{“}\delta^2\text{”}} .
\end{aligned}$$

The part of the expression labeled “$\delta^2$” is precisely the squared Mahalanobis distance from the new observation’s vector of explanatory values to the empirical distribution of the explanatory variables of the original $n$ observations. This has a beautifully intuitive interpretation: *the less the new observation resembles the original data, the more challenging it will be to predict its response value.*

Putting the terms of our derivation together,

$$\begin{aligned}
\mathbb{E}(Y_{n+1} - \widehat{Y}_{n+1})^2 &= \sigma^2 + \tfrac{\sigma^2}{n} + \tfrac{\sigma^2}{n}\delta^2 \\
&= \sigma^2[1 + \tfrac{1}{n}(1 + \delta^2)].
\end{aligned}$$

Notice that there is a limit to how well a new response can be predicted; no matter how much past data we have, this quantity remains larger than $\sigma^2$.

### 4.2.3 Bias-Variance Trade-off

When the model is correctly specified and the design matrix is full rank, the least-squares coefficients are *unbiased* estimators for the true coefficients. That means that they are *good* estimators, right? Not necessarily.

The *bias-variance trade-off* is easily one of the single most important principles in statistics. The bias-variance decomposition (Theorem 3.1) says that the expected squared distance of an estimator from the true value is the squared distance of its expectation from that value (squared bias) plus the expected squared distance from its own expectation (variance). It is naive to assume that unbiasedness is automatically a desirable property in an estimator. The objective should instead be to balance squared bias and variance such that their sum is about as small as possible.

To demonstrate this idea, we will explore how it works for the least-squares coefficients of the explanatory variables $\hat{\boldsymbol{\beta}}$ in multiple linear regression.[3]

**Exercise 4.17** Suppose $Y_1, \ldots, Y_n$ are uncorrelated and all have the same variance $\sigma^2$. Let $\mathbf{x}_1, \ldots, \mathbf{x}_n \in \mathbb{R}^m$ be $n$ observations of $m$ explanatory variables, and assume their empirical covariance matrix $\Sigma$ has full rank. Let $\hat{\boldsymbol{\beta}} = (\hat{\beta}_1, \ldots, \hat{\beta}_m)$ be the coefficients of the explanatory variables in the least-squares hyperplane $y = \hat{\alpha} + \hat{\beta}_1(x^{(1)} - \bar{x}^{(1)}) + \ldots + \hat{\beta}_m(x^{(m)} - \bar{x}^{(m)})$. Find $\mathbb{E}\|\hat{\boldsymbol{\beta}} - \mathbb{E}\hat{\boldsymbol{\beta}}\|^2$ in terms of $\sigma^2$, $n$, and the eigenvalues of $\Sigma$.

*Solution* The "variance" of any random vector is the trace of its covariance matrix (Exercise 3.25).

$$\begin{aligned}\mathbb{E}\|\hat{\boldsymbol{\beta}} - \mathbb{E}\hat{\boldsymbol{\beta}}\|^2 &= \operatorname{tr}\operatorname{cov}\hat{\boldsymbol{\beta}} \\ &= \tfrac{\sigma^2}{n}\operatorname{tr}\Sigma^{-1} \\ &= \tfrac{\sigma^2}{n}(\lambda_1^{-1} + \ldots + \lambda_m^{-1}),\end{aligned}$$

where $\lambda_1, \ldots, \lambda_m$ are the eigenvalues of $\Sigma$. ♦

One method for decreasing the variance of an estimator is called *shrinkage*. With this approach, the estimator is "pulled" toward a non-random vector (often **0**) as in Exercise 3.19. Based on that exercise, when the model is correctly specified, the optimal estimator of the form $a\hat{\boldsymbol{\beta}}$ has

$$a = \frac{\|\boldsymbol{\beta}\|^2}{\|\boldsymbol{\beta}\|^2 + (\sigma^2/n)(\lambda_1^{-1} + \ldots + \lambda_m^{-1})}. \tag{4.1}$$

We see that there is always *some* amount of shrinkage toward **0** that would improve the quality of estimation, but that is not necessarily very helpful because we do not know $\sigma^2$ and $\|\boldsymbol{\beta}\|$. However, the error variance can be estimated as in Sect. 4.2.1. And $\|\boldsymbol{\beta}\|^2$ can be estimated by $\|\hat{\boldsymbol{\beta}}\|^2$. If these estimates are substituted into Eq. 4.1, does the resulting shrinkage estimator perform better than the original $\hat{\boldsymbol{\beta}}$? It is something to think about.[4]

**Exercise 4.18** Based on Exercise 4.12, the Gauss–Markov theorem implies that the least-squares coefficient vector has the smallest possible expected squared estimation error among all random vectors that are both linear functions of the response and unbiased for its expectation. However, Eq. 4.1 identified $a < 1$ such that $a$ times the least-squares coefficients of the explanatory variables has

[3] Realize that the bias-variance trade-off plays a role in *prediction* as well via estimation of the new observation's expectation.

[4] There are some surprising findings in this vein, especially *James–Stein estimators* (see e.g. Young and Smith 2005, Sec 3.4).

smaller expected squared estimation error than the least-squares coefficient vector do; explain why this does not contradict the Gauss–Markov theorem.

*Solution* Let us check the conditions of the Gauss–Markov theorem. It applies to linear functions of the response $\mathbf{Y}$ that are unbiased for $\mathbb{E}\mathbf{Y}$. Because $\hat{\boldsymbol{\beta}}$ is linear in $\mathbf{Y}$, so is $a\hat{\boldsymbol{\beta}}$. However, it is *biased*; its expectation is $a\boldsymbol{\beta} \neq \boldsymbol{\beta}$, so Gauss–Markov does not apply. ♦

Another important way to decrease variance is to use a *simpler* model for regression. Let the explanatory variables $\mathbf{x}^{(1)}, \ldots, \mathbf{x}^{(m)}$ be the columns of $\mathbb{X}$, let $\widetilde{\mathbb{X}}$ denote its centered version, and suppose that its empirical covariance matrix $\Sigma$ is full rank. Assume

$$\mathbf{Y} = \alpha\mathbf{1} + \widetilde{\mathbb{X}}\boldsymbol{\beta} + \boldsymbol{\epsilon}$$

with $\mathbb{E}\boldsymbol{\epsilon} = \mathbf{0}$. Consider doing multiple linear regression while *omitting* the final explanatory variable. In order to figure out how much this would increase the squared bias, let us work out the expectation of these alternative least-squares coefficients.

With $\mathbb{X}_0 \in \mathbb{R}^{n\times(m-1)}$ as the matrix whose columns are $\mathbf{x}^{(1)}, \ldots, \mathbf{x}^{(m-1)}$, its centered version denoted $\widetilde{\mathbb{X}}_0$, and its empirical covariance matrix $\Sigma_0 \in \mathbb{R}^{(m-1)\times(m-1)}$, the alternative least-squares coefficient vector $\check{\boldsymbol{\beta}}$ has

$$\begin{bmatrix} \check{\beta}_1 \\ \vdots \\ \check{\beta}_{m-1} \end{bmatrix} = \Sigma_0^{-1}\frac{1}{n}\widetilde{\mathbb{X}}_0^T\mathbf{Y}.$$

Its expectation is

$$\mathbb{E}\begin{bmatrix} \check{\beta}_1 \\ \vdots \\ \check{\beta}_{m-1} \end{bmatrix} = \Sigma_0^{-1}\frac{1}{n}\widetilde{\mathbb{X}}_0^T \underbrace{\mathbb{E}\mathbf{Y}}_{\alpha\mathbf{1}+\widetilde{\mathbb{X}}\boldsymbol{\beta}}$$
$$= \Sigma_0^{-1}\frac{1}{n}\widetilde{\mathbb{X}}_0^T\widetilde{\mathbb{X}}\boldsymbol{\beta}.$$

The first $m-1$ columns of $\frac{1}{n}\widetilde{\mathbb{X}}_0^T\widetilde{\mathbb{X}}$ are the same as $\Sigma_0$, while the final column is $(\sigma_{\mathbf{x}^{(1)},\mathbf{x}^{(m)}}, \ldots, \sigma_{\mathbf{x}^{(m-1)},\mathbf{x}^{(m)}})$. When this matrix is multiplied by $\Sigma_0^{-1}$ on the left, the product looks like an identity matrix with an extra column on the right that is

$$\begin{bmatrix} r_1 \\ \vdots \\ r_{m-1} \end{bmatrix} := \Sigma_0\begin{bmatrix} \sigma_{\mathbf{x}^{(1)},\mathbf{x}^{(m)}} \\ \vdots \\ \sigma_{\mathbf{x}^{(m-1)},\mathbf{x}^{(m)})} \end{bmatrix}.$$

The astute reader will recognize these entries as the least-squares coefficients of $\mathbf{x}^{(1)}, \ldots, \mathbf{x}^{(m-1)}$ when they are used as the explanatory variables to predict $\mathbf{x}^{(m)}$ via multiple linear regression (Theorem 2.3). Finally, multiplying the matrix by $\boldsymbol{\beta}$ we find a simple expression for the expectation of the alternative least-squares coefficients:

$$\mathbb{E}\check{\beta}_j = \beta_j + r_j \beta_m$$

for $j \in \{1, \ldots, m-1\}$. Implicitly the coefficient $\beta^{(m)}$ is estimated by $\check{\beta}_m = 0$ which has expectation 0. The difference $\mathbb{E}\check{\boldsymbol{\beta}} - \boldsymbol{\beta}$ simplifies to $\beta_m(r_1, \ldots, r_{m-1}, -1)$, so the "squared bias" is

$$\|\mathbb{E}\check{\boldsymbol{\beta}} - \boldsymbol{\beta}\|^2 = \beta_m^2 \left(1 + \sum_i r_i^2\right).$$

If $\operatorname{cov} \boldsymbol{\epsilon} = \sigma^2 \mathbb{I}$, the "variance" of $\check{\boldsymbol{\beta}}$ is

$$\mathbb{E}\|\check{\boldsymbol{\beta}} - \mathbb{E}\check{\boldsymbol{\beta}}\|^2 = \tfrac{\sigma^2}{n}(\xi_1^{-1} + \ldots + \xi_{m-1}^{-1}),$$

where $\xi_1, \ldots, \xi_{m-1}$ are the eigenvalues of $\Sigma_0$.

It is easy to see that $\check{\boldsymbol{\beta}}$ has a smaller variance that $\hat{\boldsymbol{\beta}}$ in the case of only two explanatory variables. The variance of $\mathbf{x}^{(1)}$ cannot be smaller than $\lambda_2$, the smallest eigenvalue of the empirical covariance matrix, so

$$\begin{aligned}
\mathbb{E}\|\check{\boldsymbol{\beta}} - \boldsymbol{\beta}\|^2 &= \tfrac{\sigma^2}{n} \tfrac{1}{\sigma^2_{\mathbf{x}^{(1)}}} \\
&\leq \tfrac{\sigma^2}{n} \tfrac{1}{\lambda_2} \\
&< \tfrac{\sigma^2}{n} (\tfrac{1}{\lambda_1} + \tfrac{1}{\lambda_2}) \\
&= \mathbb{E}\|\hat{\boldsymbol{\beta}} - \boldsymbol{\beta}\|^2.
\end{aligned}$$

The effect on squared bias and variance is much neater if the *principal components* of $\mathbb{X}$ were instead used as the variables. The principal components are all uncorrelated with each other, so the squared bias would simplify to the omitted PC variable's true coefficient, while the decrease in variance simplifies to $\sigma^2/n$ times the reciprocal of the variance of the omitted PC variable.

Note that the $(1/n)$ factor means the variance term will tend to carry less weight in the bias-variance trade-off as the sample size increases. In other words, the larger your sample size, the more willing you should be to prefer complex models. This is typical of the bias-variance trade-off. In fact, we can see the same phenomenon in our shrinkage example: as the sample size increases, the optimal factor multiplying $\hat{\beta}$ approaches 1.

To reiterate, the key to high-quality estimation or prediction is to appropriately balance variance and bias; using a *perfectly accurate* model is not so important.

Ideally *many* models should be under consideration so that the form of the relationship between the response and explanatory variables can be revealed by the data instead of being imposed by the practitioner. However, if the only criterion for picking a model and parameter values is minimizing the sum of squared residuals, your estimation procedure will tend to have such a large variance that it will perform poorly. Various *model selection* techniques have been devised to help prevent choosing a model with excess variance. For many of them, the model and parameter values come from minimizing an objective function that has two terms: one that quantifies the model's misfit from the data (roughly indicating the estimator's bias) such as sum of squared residuals and another that quantifies the model's "complexity" (roughly indicating the estimator's variance). The hope that a good balance between complexity and misfit corresponds to a good balance between bias and variance. Model selection is an essential topic for any statistician, but it is beyond the scope of this text.

Next, we will study the Normal distribution in Chap. 5 so that in Chap. 6 we can add a modeling assumption that the errors are Normal. As you will see, this assumption opens the door to a tremendous variety of new opportunities for statistical inference.

# Chapter 5
# Background: Normality

The family of Normal distributions plays a key role in the theory of probability and statistics. According to the familiar *Central Limit Theorem*, the distribution of an average of iid random variables (with finite variance) tends toward Normality (Pollard 2002, Thm 7.21). In fact, more advanced versions of the theorem do not require the random variables to be iid, as long as they are not *too* dependent or *too* disparate in their scales (e.g. Pollard 2002, Thm 8.14). We see this Central Limit phenomenon play out in the real world when we observe "bell-shaped" histograms of measurements in a wide range of contexts. The prevalence of approximate Normality in the world makes Normal distributions a natural part of statistical modeling. Fortunately, the Normal family is also mathematically convenient for analyzing estimation procedures for these models.

In this chapter, we'll work out properties of the Normal family that will gain us much when we finally assume that the linear model's errors are Normal in Chap. 6.

## 5.1 Standard Normal

**Definition (Standard Normal Distribution)** A random variable has the **standard Normal distribution**, denoted $N(0, 1)$, if it has probability density function

$$f(z) = \frac{1}{\sqrt{2\pi}} e^{-z^2/2}.$$

An $\mathbb{R}^n$-valued random vector has the **standard Normal distribution** on $\mathbb{R}^n$, denoted $N(\mathbf{0}, \mathbb{I})$, if its coordinates are iid $N(0, 1)$ random variables

**Exercise 5.1** Find the probability density function for a standard Normal random vector on $\mathbb{R}^n$.



W. D. Brinda, *Visualizing Linear Models*,
https://doi.org/10.1007/978-3-030-64167-2_5

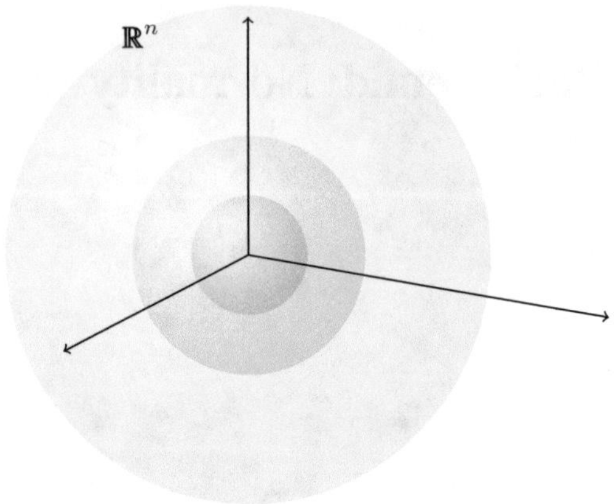


**Fig. 5.1** If its coordinates $Z_1, \ldots, Z_n$ are iid $N(0, 1)$, then $\mathbf{Z} \sim N(\mathbf{0}, \mathbb{I}_n)$. The probability density function of this distribution is spherically symmetric about $\mathbf{0}$. A spherical symmetric function has spheres as its level sets

*Solution* Let $\mathbf{Z}$ be an $\mathbb{R}^n$-valued standard Normal random vector. By independence, its pdf equals the product of the individual pdfs of its coordinates $(Z_1, \ldots, Z_n)$.

$$\begin{aligned} f(\mathbf{z}) &= \prod_i \frac{1}{\sqrt{2\pi}} e^{-z_i^2/2} \\ &= \frac{1}{(2\pi)^{n/2}} e^{-(z_1^2+\ldots+z_n^2)/2} \\ &= \frac{1}{(2\pi)^{n/2}} e^{-\|\mathbf{z}\|^2/2} \end{aligned}$$

♦

The standard Normal pdf decreases as you travel away from the origin. Importantly, *the density only depends on the distance from the origin*: a point of distance $t$ from the origin has density $\frac{1}{(2\pi)^{n/2}} e^{-t^2/2}$ *no matter which direction the point is in*. Spherical level sets for its pdf are drawn in Fig. 5.1 from a generic three-dimensional perspective. Much of the analysis in Chap. 6 will stem from this *spherical symmetry* of the $N(\mathbf{0}, \mathbb{I})$ distribution.

Suppose[1] $\mathbf{Z} \sim N(\mathbf{0}, \mathbb{I}_n)$, and let $\mathbf{u}_1, \ldots, \mathbf{u}_n$ be an orthonormal basis for $\mathbb{R}^n$. Consider the distribution of its vector of coordinates $(\langle \mathbf{Z}, \mathbf{u}_1\rangle, \ldots, \langle \mathbf{Z}, \mathbf{u}_n\rangle)$. The pdf of $\mathbf{Z}$ can be rewritten in terms of these coordinates by Parseval's identity

$$\begin{aligned} f(\mathbf{z}) &= \frac{1}{(2\pi)^{n/2}} e^{-\|\mathbf{z}\|^2/2} \\ &= \frac{1}{(2\pi)^{n/2}} e^{-(\langle \mathbf{z},\mathbf{u}_1\rangle^2+\ldots+\langle \mathbf{z},\mathbf{u}_n\rangle^2)/2} \\ &= \prod_i \frac{1}{\sqrt{2\pi}} e^{-\langle \mathbf{z},\mathbf{u}_i\rangle^2/2}. \end{aligned}$$

[1]The notation $\mathbf{Z} \sim N(\mathbf{0}, \mathbb{I}_n)$ is short for "$\mathbf{Z}$ has the distribution $N(\mathbf{0}, \mathbb{I}_n)$."

The density of the alternative coordinates $(\langle \mathbf{Z}, \mathbf{u}_1 \rangle, \ldots, \langle \mathbf{Z}, \mathbf{u}_n \rangle)$ is exactly the same[2] as the density of the standard coordinates $(Z_1, \ldots, Z_n)$, so they must have the same distribution. *With any orthonormal basis, the coordinates of a standard Normal random vector are iid standard Normal random variables.* This observation should feel intuitive due to the nature of spherical symmetry; the density "looks the same" no matter how you orient the orthogonal axes.

With $\mathbf{Z} \sim N(\mathbf{0}, \mathbb{I})$, let us think about the orthogonal projection of $\mathbf{Z}$ onto a subspace $\mathcal{S}$. Consider an alternative orthonormal basis $\mathbf{u}_1, \ldots, \mathbf{u}_n$ which has its first $d := \dim(\mathcal{S})$ vectors spanning $\mathcal{S}$ and its remaining $n - d$ vectors (necessarily) orthogonal to $\mathcal{S}$ (see Fig. 5.2). As we learned in Exercise 1.42, the orthogonal projection of $\mathbf{Z}$ onto $\mathcal{S}$ is $\widehat{\mathbf{Z}} := \langle \mathbf{Z}, \mathbf{u}_1 \rangle \mathbf{u}_1 + \ldots + \langle \mathbf{Z}, \mathbf{u}_d \rangle \mathbf{u}_d$. The random vector $\mathbf{Z} - \widehat{\mathbf{Z}}$ is the orthogonal projection onto the orthogonal complement of $\mathcal{S}$, and it is equal to $\langle \mathbf{Z}, \mathbf{u}_{d+1} \rangle \mathbf{u}_{d+1} + \ldots + \langle \mathbf{Z}, \mathbf{u}_n \rangle \mathbf{u}_n$. Notice that $\widehat{\mathbf{Z}}$ is a function of only the first $d$ coordinates (with respect to the alternative basis) and $\mathbf{Z} - \widehat{\mathbf{Z}}$ is a function of only the remaining coordinates. Because the coordinates are independent, we can conclude that the orthogonal projections $\widehat{\mathbf{Z}}$ and $\mathbf{Z} - \widehat{\mathbf{Z}}$ are independent of each other.

It is important to understand how this conclusion extends when there are more subspaces in question. If $\mathcal{S}_1, \ldots, \mathcal{S}_m$ are subspaces that are all orthonormal to

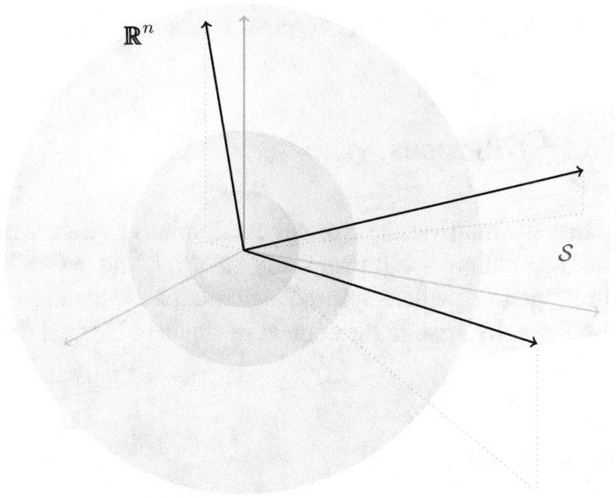


**Fig. 5.2** What if $\mathbf{Z}$ is represented by another choice of orthonormal basis vectors? By spherical symmetry, it is intuitive that the distribution of the coordinate random variables must be the same regardless of the orientation of the axes. In particular, we can consider an orthonormal basis in which the first $\dim(\mathcal{S})$ axes are chosen from $\mathcal{S}$, and the remaining $n-\dim(\mathcal{S})$ are (necessarily) chosen orthogonally to $\mathcal{S}$. In the figure, the original axes have been grayed out, and new axes are drawn

[2]When changing variables from one rectangular coordinate system to another, one also needs to multiply by the ratio of the volumes of the differential elements. In this case, that ratio is 1 as the volumes are the same.

each other, then the orthogonal projections of $\mathbf{Z}$ onto those subspaces can be represented as functions of distinct coordinates with respect to some orthonormal basis. Therefore, *the orthogonal projections of a standard Normal random vector onto orthogonal subspaces are independent of each other*.

In general, we do not need to explicitly construct an alternative coordinate system. The alternative coordinate system is merely a conceptual tool that enables us to understand the distributions of a number of statistics that will appear in Chap. 6. We'll study those distributions in the remainder of this chapter.

Before moving on, however, let us observe one more foundational fact. Let $a_1, a_2 \in \mathbb{R}$ and $Z_1, Z_2 \overset{iid}{\sim} N(0, 1)$. Then

$$\frac{a_1 Z_1 + a_2 Z_2}{\sqrt{a_1^2 + a_2^2}} \sim N(0, 1).$$

To see why, think of the standard Normal random vector $\mathbf{Z} := (Z_1, Z_2)$. Based on the preceding discussion, its coordinate with respect to the unit vector $\mathbf{u} := \frac{1}{\sqrt{a_1^2+a_2^2}}(a_1, a_2)$ must be standard Normal. This coordinate is exactly $\langle \mathbf{Z}, \mathbf{u} \rangle = \frac{a_1 Z_1 + a_2 Z_2}{\sqrt{a_1^2+a_2^2}}$. More generally, any linear combination of iid standard Normal random variables, divided by the norm of the coefficient vector, is standard Normal.

## 5.2 Derived Distributions

This section defines the multivariate Normal distributions as transformations of the standard Normal distribution. We'll then study the $\chi^2$, $t$, and $f$ families along with their "non-central" generalizations. Associated exercises demonstrate how these distributions also naturally arise in the context of standard Normal draws.

### *5.2.1 Normal Family*

**Definition (Normal)** Let $\mathbf{Z} \sim N(\mathbf{0}, \mathbb{I}_m)$. For any $\mathbb{M} \in \mathbb{R}^{n \times m}$ and $\boldsymbol{\mu} \in \mathbb{R}^n$, the distribution of $\mathbb{M}\mathbf{Z} + \boldsymbol{\mu}$ is called a **Normal distribution** and is characterized by its expectation and covariance.

With $\mathbf{Z} \sim N(\mathbf{0}, \mathbb{I}_m)$, the expectation and covariance of $\mathbf{Y} := \mathbb{M}\mathbf{Z} + \boldsymbol{\mu}$ can be calculated (via Exercises 3.9 and 3.23) to be $\boldsymbol{\mu}$ and $\mathbb{M}\mathbb{M}^T$.

Alternatively, we learned from Exercise 3.33 and the text's subsequent discussion that a standardized random vector can be transformed to have any desired expectation and covariance by using just the right *affine* mapping. If $\mathbf{Z} \sim N(\mathbf{0}, \mathbb{I})$, then

$\mathbb{C}^{1/2}\mathbf{Z}+\boldsymbol{\mu} \sim N(\boldsymbol{\mu}, \mathbb{C})$. The idea of standardizing and "un-standardizing" (Sect. 3.5) while retaining Normality will be a recurring trick in the coming exercises.

**Exercise 5.2** For standard Normal random vectors $\mathbf{Z}_1$ and $\mathbf{Z}_2$, suppose $\mathbb{M}_1\mathbf{Z}_1$ and $\mathbb{M}_2\mathbf{Z}_2$ have the same covariance. Show that they have the same distribution.

**Exercise 5.3** Show that if $\mathbf{X}$ is a Normal random vector, then so is $\mathbb{M}\mathbf{X} + \mathbf{v}$ where $\mathbb{M}$ is a real matrix and $\mathbf{v}$ is a vector.

*Solution* With $\boldsymbol{\mu}$ and $\mathbb{C}$ representing the expectation and covariance of $\mathbf{X}$, the transformed random vector is

$$\begin{aligned}\mathbb{M}\mathbf{X} + \mathbf{v} &= \mathbb{M}(\mathbb{C}^{1/2}\mathbf{Z} + \boldsymbol{\mu}) + \mathbf{v} \\ &= [\mathbb{M}\mathbb{C}^{1/2}]\mathbf{Z} + [\mathbb{M}\boldsymbol{\mu} + \mathbf{v}]\end{aligned}$$

with $\mathbf{Z}$ standard Normal. This fits the definition of a Normal random vector. ♦

**Exercise 5.4** Show that if $\mathbf{X}_1$ and $\mathbf{X}_2$ are Normal random vectors, then so is $\mathbf{X}_1 + \mathbf{X}_2$.

**Exercise 5.5** If two random variables are *multivariate* Normal and are uncorrelated with each other, then they are independent; one can verify that their joint density factors into a product of their marginal densities. However, without *multivariate* Normality, uncorrelated does not necessarily imply independent. Construct a pair of Normal random variables that are uncorrelated but not independent.

### 5.2.2 $\chi^2$-Distributions

**Definition** **($\chi^2$-Distribution)** If $\mathbf{Z} = (Z_1, \ldots, Z_k)$ is a standard Normal random vector, the distribution of $\|\mathbf{Z}\|^2 = Z_1 + \ldots + Z_k$ is called the **$\chi^2$-distribution** with $k$ degrees of freedom, denoted $\chi_k^2$.

**Exercise 5.6** Find the expectation of $W \sim \chi_k^2$.

*Solution* $W$ can be represented as the squared norm of a standard Normal random vector. Its expectation is the same as the expected squared norm of *any* standardized random vector $\mathbf{Z}$ on $\mathbb{R}^k$:

$$\begin{aligned}\mathbb{E}\|\mathbf{Z}\|^2 &= \mathbb{E}(Z_1^2 + \ldots + Z_k^2) \\ &= \mathbb{E}Z_1^2 + \ldots + \mathbb{E}Z_k^2\end{aligned}$$

$$= \underbrace{\operatorname{var} Z_1}_{1} + \ldots + \underbrace{\operatorname{var} Z_k}_{1}$$
$$= k.$$

♦

**Exercise 5.7** If $\mathbf{Y} \sim N(\boldsymbol{\mu}, \mathbb{C})$ is an $\mathbb{R}^n$-valued random vector, what is the distribution of the squared Mahalanobis distance of $\mathbf{Y}$ from its own distribution?

*Solution* Allow for degenerate distributions by using the approach described at the end of Sect. 3.5. Let $\mathbf{Z} := \mathbb{C}^{-1/2}(\mathbf{Y} - \boldsymbol{\mu}) \sim N(\mathbf{0}, \mathbb{I})$ represent the standardized version in $\mathbb{R}^{\operatorname{rank} \mathbb{C}}$. The squared Mahalanobis distance from $\mathbf{Y}$ to $N(\boldsymbol{\mu}, \mathbb{C})$ is

$$\|\mathbb{C}^{-1/2}[\mathbf{Y} - \boldsymbol{\mu}]\|^2 = \|\mathbb{C}^{-1/2}[(\mathbb{C}^{1/2}\mathbf{Z} + \boldsymbol{\mu}) - \boldsymbol{\mu}]\|^2$$
$$= \|\mathbf{Z}\|^2$$
$$\sim \chi^2_{\operatorname{rank} \mathbb{C}}.$$

♦

**Exercise 5.8** Let $\mathbf{Z}$ be an $\mathbb{R}^n$-valued random vector with the standard Normal distribution, and let $\mathbb{H}$ be an orthogonal projection matrix. Find the distribution of $\|\mathbb{H}\mathbf{Z}\|^2$.

*Solution* Let $\mathbf{u}_1, \ldots, \mathbf{u}_n$ be an orthonormal basis with $\mathbf{u}_1, \ldots, \mathbf{u}_{\operatorname{rank} \mathbb{H}}$ spanning the space that $\mathbb{H}$ projects onto. Because the orthogonal projection is

$$\mathbb{H}\mathbf{Z} = \langle \mathbf{Z}, \mathbf{u}_1 \rangle \mathbf{u}_1 + \ldots + \langle \mathbf{Z}, \mathbf{u}_{\operatorname{rank} \mathbb{H}} \rangle \mathbf{u}_{\operatorname{rank} \mathbb{H}}$$

its squared length is the sum of its squared coordinates

$$\|\mathbb{H}\mathbf{Z}\|^2 = \langle \mathbf{Z}, \mathbf{u}_1 \rangle^2 + \ldots + \langle \mathbf{Z}, \mathbf{u}_{\operatorname{rank} \mathbb{H}} \rangle^2.$$

These coordinates are independent standard Normal random variables, according to the discussion in Sect. 5.1, so their sum of squares has distribution $\chi^2_{\operatorname{rank} \mathbb{H}}$. ♦

Now we'll generalize this family to include the squared lengths of *translated* standard Normal random vectors. By spherical symmetry, it is clear that the *direction* of translation will not matter; only the distance will.

**Definition (Non-central $\chi^2$-Distribution)** If $\mathbf{Z} = (Z_1, \ldots, Z_k)$ is a standard Normal random vector and $\mathbf{v} \in \mathbb{R}^k$, then the distribution of $\|\mathbf{v} + \mathbf{Z}\|^2$ is called the **non-central $\chi^2$-distribution** with $k$ degrees of freedom and non-centrality parameter $\|\mathbf{v}\|^2$, denoted $\chi^2_{k, \|\mathbf{v}\|^2}$.

### 5.2.3 t-Distributions

As we'll see in Chap. 6, many statistics depend on a scaling factor that is *unknown* even after specifying a null hypothesis of interest. However, in many cases, statisticians have cleverly devised a ratio of two independent statistics that are both proportional to that factor, making it cancel out; the distribution of the resulting ratio does not depend on the unknown scaling factor, allowing us to know its distribution exactly.

**Definition** (**$t$-Distribution**) If $Z \sim N(0, 1)$ and $V \sim \chi_k^2$ are independent of each other, then the distribution of the ratio $\frac{Z}{\sqrt{V/k}}$ is called the **t-distribution** with $k$ degrees of freedom, denoted $t_k$.

Notice that in the above definition, $V/k$ can be represented as an average $\frac{1}{k}(Z_1^2 + \ldots + Z_k^2)$ with $Z_1, \ldots, Z_k \overset{iid}{\sim} N(0, 1)$. By the law of large numbers (Pollard 2002, Thm 4.3), $V/k$ becomes increasingly concentrated near its expectation of 1 as $k$ increases, so the distribution of $\frac{Z}{\sqrt{V/k}}$ increasingly resembles the distribution of $Z \sim N(0, 1)$.

**Exercise 5.9** Let $\boldsymbol{\epsilon} \sim N(\mathbf{0}, \sigma^2\mathbb{I})$. If $\mathbb{H}$ is an orthogonal projection matrix and $\mathbf{u}$ is a unit vector orthogonal to $C(\mathbb{H})$, find the distribution of

$$\frac{\langle \boldsymbol{\epsilon}, \mathbf{u} \rangle}{\|\mathbb{H}\boldsymbol{\epsilon}\| / \sqrt{\operatorname{rank} \mathbb{H}}}.$$

*Solution* First, we'll divide the numerator and the denominator by $\sigma$ to connect this ratio to the standard Normal random vector $\boldsymbol{\epsilon}/\sigma$.

$$\frac{\langle \boldsymbol{\epsilon}, \mathbf{u} \rangle}{\|\mathbb{H}\boldsymbol{\epsilon}\| / \sqrt{\operatorname{rank} \mathbb{H}}} = \frac{\langle \boldsymbol{\epsilon}/\sigma, \mathbf{u} \rangle}{\|\mathbb{H}(\boldsymbol{\epsilon}/\sigma)\| / \sqrt{\operatorname{rank} \mathbb{H}}}$$

Let $\mathbf{u}_1, \ldots, \mathbf{u}_n$ be an orthonormal basis with $\mathbf{u}_1, \ldots, \mathbf{u}_{\operatorname{rank} \mathbb{H}}$ spanning $C(\mathbb{H})$ and $\mathbf{u}_{\operatorname{rank} \mathbb{H}+1}$ equal to $\mathbf{u}$. The numerator is simply the coordinate of $\boldsymbol{\epsilon}/\sigma$ with respect to $\mathbf{u}$, so it is a standard Normal random variable. From Exercise 5.8, $\|\mathbb{H}(\boldsymbol{\epsilon}/\sigma)\|^2 \sim \chi^2_{\operatorname{rank} \mathbb{H}}$. Because the numerator and the denominator are functions of distinct coordinates, they are independent of each other, so the random variable has the $t_{\operatorname{rank} \mathbb{H}}$ distribution. ♦

**Definition** (**Non-central $t$-Distribution**) If $Z \sim N(0, 1)$ and $V \sim \chi_k^2$ are independent of each other, and $a \in \mathbb{R}$, then the distribution of the ratio $\frac{Z+a}{\sqrt{V/k}}$ is called the **non-central t-distribution** with $k$ degrees of freedom and non-centrality parameter $a$, denoted $t_{k,a}$.

**Exercise 5.10** Let $\boldsymbol{\epsilon} \sim N(\mathbf{0}, \sigma^2\mathbb{I})$. If $\mathbb{H}$ is an orthogonal projection matrix and $\mathbf{u}$ is a unit vector orthogonal to $C(\mathbb{H})$, and $a \in \mathbb{R}$, find the distribution of

$$\frac{a + \langle \boldsymbol{\epsilon}, \mathbf{u} \rangle}{\|\mathbb{H}\boldsymbol{\epsilon}\| / \sqrt{\operatorname{rank} \mathbb{H}}}.$$

*Solution* First, we'll divide the numerator and the denominator by $\sigma$.

$$\frac{a + \langle \boldsymbol{\epsilon}, \mathbf{u} \rangle}{\|\mathbb{H}\boldsymbol{\epsilon}\| / \sqrt{\operatorname{rank} \mathbb{H}}} = \frac{a/\sigma + \langle (\boldsymbol{\epsilon}/\sigma), \mathbf{u} \rangle}{\|\mathbb{H}(\boldsymbol{\epsilon}/\sigma)\| / \sqrt{\operatorname{rank} \mathbb{H}}}$$

As in Exercise 5.9, the second term in the numerator is standard Normal, the denominator is $\chi^2_{\operatorname{rank} \mathbb{H}}$ divided by its degrees of freedom, and the numerator and denominator are independent. By the definition of non-central $t$-distributions, the ratio's distribution is $t_{\operatorname{rank} \mathbb{H}, a/\sigma}$. ♦

### 5.2.4 *f*-Distributions

**Definition** (*f*-**Distribution**) If $V \sim \chi^2_k$ and $W \sim \chi^2_m$ are independent of each other, then the distribution of the ratio $\frac{V/k}{W/m}$ is called the **f-distribution** with $k$ numerator degrees of freedom and $m$ denominator degrees of freedom, denoted $f_{k,m}$.

**Exercise 5.11** Let $T \sim t_k$. What is the distribution of $T^2$?

*Solution* From the definition of $t_k$, we can represent $T$ using independent $Z \sim N(0, 1)$ and $V \sim \chi^2_k$.

$$\begin{aligned} T^2 &= \left( \frac{Z}{\sqrt{V/k}} \right)^2 \\ &= \frac{Z^2/1}{V/k} \end{aligned}$$

Because $Z^2 \sim \chi^2_1$, this expression matches the definition of the $f_{1,k}$ distribution. ♦

**Exercise 5.12** Let $\boldsymbol{\epsilon} \sim N(\mathbf{0}, \sigma^2\mathbb{I})$, and let $\mathbb{H}_1$ and $\mathbb{H}_2$ be orthogonal projection matrices onto two subspaces that are orthogonal to each other. Find the distribution of $\frac{\|\mathbb{H}_1\boldsymbol{\epsilon}\|^2/\operatorname{rank}\mathbb{H}_1}{\|\mathbb{H}_2\boldsymbol{\epsilon}\|^2/\operatorname{rank}\mathbb{H}_2}$.

*Solution* We can divide both the numerator and the denominator by $\sigma^2$ to produce random variables whose distributions we know from Exercise 5.8.

$$\frac{\|\mathbb{H}_1\boldsymbol{\epsilon}\|^2/\operatorname{rank}\mathbb{H}_1}{\|\mathbb{H}_2\boldsymbol{\epsilon}\|^2/\operatorname{rank}\mathbb{H}_2} = \frac{\|\mathbb{H}_1(\boldsymbol{\epsilon}/\sigma)\|^2/\operatorname{rank}\mathbb{H}_1}{\|\mathbb{H}_2(\boldsymbol{\epsilon}/\sigma)\|^2/\operatorname{rank}\mathbb{H}_2}$$

The numerator is a $\chi^2_{\text{rank}\,\mathbb{H}_1}$-distributed random variable divided by its degrees of freedom, while the denominator is a $\chi^2_{\text{rank}\,\mathbb{H}_2}$-distributed random variable divided by its degrees of freedom. Because the subspaces are orthogonal, we know that the two orthogonal projections are independent of each other, allowing us to conclude that the ratio matches the definition of $f_{\text{rank}\,\mathbb{H}_1,\text{rank}\,\mathbb{H}_2}$. ♦

**Definition (Non-central $f$-Distribution)** If $V \sim \chi^2_{k,a}$ and $W \sim \chi^2_m$ are independent of each other, then the distribution of the ratio $\frac{V/k}{W/m}$ is called the **non-central f-distribution** with $k$ numerator degrees of freedom, $m$ denominator degrees of freedom, and non-centrality parameter $a$, denoted $f_{k,m,a}$.

**Exercise 5.13** Let $\boldsymbol{\epsilon} \sim N(\mathbf{0}, \sigma^2\mathbb{I})$, and let $\mathbb{H}_1$ and $\mathbb{H}_2$ be orthogonal projection matrices onto two subspaces that are orthogonal to each other. Find the distribution of $\frac{\|\mathbf{v}+\mathbb{H}_1\boldsymbol{\epsilon}\|^2/\text{rank}\,\mathbb{H}_1}{\|\mathbb{H}_2\boldsymbol{\epsilon}\|^2/\text{rank}\,\mathbb{H}_2}$, where $\mathbf{v}$ is a non-random vector.

*Solution* Divide both the numerator and the denominator by $\sigma^2$.

$$\frac{\|\mathbf{v}+\mathbb{H}_1\boldsymbol{\epsilon}\|^2/\text{rank}\,\mathbb{H}_1}{\|\mathbb{H}_2\boldsymbol{\epsilon}\|^2/\text{rank}\,\mathbb{H}_2} = \frac{\|\frac{1}{\sigma}\mathbf{v}+\mathbb{H}_1(\boldsymbol{\epsilon}/\sigma)\|^2/\text{rank}\,\mathbb{H}_1}{\|\mathbb{H}_2(\boldsymbol{\epsilon}/\sigma)\|^2/\text{rank}\,\mathbb{H}_2}$$

As in Exercise 5.12, the denominator is $\chi^2_{\text{rank}\,\mathbb{H}_2}$-distributed and is independent of the numerator. This time the numerator is a non-central $\chi^2$ random variable divided by its degrees of freedom with non-centrality parameter $\|\frac{1}{\sigma}\mathbf{v}\|^2 = \|\mathbf{v}\|^2/\sigma^2$. Thus the ratio's distribution matches the definition of $f_{\text{rank}\,\mathbb{H}_1,\text{rank}\,\mathbb{H}_2,\|\mathbf{v}\|^2/\sigma^2}$. ♦

Now that we have developed intuition and tools for Normal random vectors, we can put them into practice in the context of linear modeling with the assumption that the errors are Normal.

# Chapter 6
# Normal Errors

Throughout this chapter, we will continue analyzing the linear model from Chap. 4, but we will assume in particular that $\boldsymbol{\epsilon} \sim N(\mathbf{0}, \sigma^2\mathbb{I})$ for an unknown $\sigma^2$. The figures in Chap. 4 already indicated iid Normal errors, although the results we derived in that chapter did not require such strong assumptions about the distribution of the errors. With the new Normality assumption, our earlier results remain valid of course, but we will also be able to do a good deal more in terms of inference.

**Exercise 6.1** Let $\mathbf{x}_i$ represent the explanatory value(s) of the $i$th observation. Consider modeling the response variable by

$$Y_i = f_\theta(\mathbf{x}_i) + \epsilon_i$$

with $\epsilon_1, \ldots, \epsilon_n \overset{iid}{\sim} N(0, \sigma^2)$ and $\theta \in \Theta$ indexing a set of possible functions. (Notice that this form is far more general than the linear model with iid Normal errors.) Show that the maximum likelihood estimator for $\theta$ is precisely the parameter value that minimizes the sum of squared residuals.

*Solution* The response values have distribution $Y_i \sim N(f_\theta(\mathbf{x}_i), \sigma^2)$ and are independent of each other. Because of independence, the overall likelihood $L(\theta; \mathbf{Y})$ is the product of the individual observations' likelihoods.

$$\begin{aligned} L(\theta; \mathbf{Y}) &= \prod_{i=1}^{n} \frac{1}{\sqrt{2\pi\sigma^2}} e^{-\frac{1}{2\sigma^2}(Y_i - f_\theta(\mathbf{x}_i))^2} \\ &= \left(\frac{1}{2\pi\sigma^2}\right)^{n/2} e^{-\frac{1}{2\sigma^2}\sum_{i=1}^{n}(Y_i - f_\theta(\mathbf{x}_i))^2}. \end{aligned}$$

The parameter $\theta$ only appears in the sum of squared residuals $\sum_{i=1}^{n}(Y_i - f_\theta(\mathbf{x}_i))^2$. The smaller the sum of squared residuals is, the larger the likelihood is, so the



W. D. Brinda, *Visualizing Linear Models*,
https://doi.org/10.1007/978-3-030-64167-2_6

"least-squares parameter" is exactly the maximum likelihood estimator. Notice that this equivalence does not depend on the value of $\sigma$ and that it holds even if $\sigma$ is unknown. ♦

## 6.1 Distribution of Sums of Squares

The first observation to make is that many of the important random vectors related to least-squares linear regression are linear or affine transformations of the response $\mathbf{Y} = \mathbb{E}\mathbf{Y} + \boldsymbol{\epsilon}$ and therefore affine functions of the error vector. If $\epsilon$ is Normal, then so are these random vectors. The expectations and covariance matrices of many such random vectors were derived in Chap. 4. Thus with $\boldsymbol{\epsilon} \sim N(\mathbf{0}, \sigma^2\mathbb{I})$, we now know the exact distribution of, for example,

$$\begin{aligned}\widehat{\mathbf{Y}} &:= \mathbb{H}\mathbf{Y} \\ &\sim N(\mathbb{H}\mathbb{E}\mathbf{Y}, \sigma^2\mathbb{H}),\end{aligned}$$

where $\mathbb{H}$ is the orthogonal projection matrix onto the design matrix's column space. It is also important to realize that even if the error is not Normal or perfectly independent, the distribution of these statistics will still be approximately Normal if the sample size is large enough due to the central limit phenomenon.

**Exercise 6.2** Suppose $\mathbf{Y} = \mathbb{M}\boldsymbol{\gamma} + \boldsymbol{\epsilon}$ with error vector $\boldsymbol{\epsilon} \sim N(\mathbf{0}, \sigma^2\mathbb{I})$. If $\widehat{\mathbf{Y}}$ is the least-squares linear regression's prediction vector for design matrix $\mathbb{M}$, what is the distribution of $\|\mathbf{Y} - \widehat{\mathbf{Y}}\|^2/\sigma^2$?

*Solution* We saw in Chap. 4 that the least-squares residual vector $\mathbf{Y} - \widehat{\mathbf{Y}}$ is the orthogonal projection of $\boldsymbol{\epsilon}$ onto $C(\mathbb{M})^\perp$ which has dimension $n - \operatorname{rank}\mathbb{M}$. The standardized version $\boldsymbol{\epsilon}/\sigma$ is standard Normal, so according to Exercise 5.8,

$$\begin{aligned}\frac{\|\mathbf{Y} - \widehat{\mathbf{Y}}\|^2}{\sigma^2} &= \|(\mathbb{I} - \mathbb{H})(\boldsymbol{\epsilon}/\sigma)\|^2 \\ &\sim \chi^2_{n-\operatorname{rank}\mathbb{M}},\end{aligned}$$

where $\mathbb{H}$ represents the orthogonal projection matrix onto $C(\mathbb{M})$. ♦

Other squared norms of projections of the error vector work out likewise. As in Exercise 5.8, if $\mathbb{H}$ is *any* orthogonal projection matrix,

$$\begin{aligned}\frac{\|\mathbb{H}\boldsymbol{\epsilon}\|^2}{\sigma^2} &= \|\mathbb{H}(\boldsymbol{\epsilon}/\sigma)\|^2 \\ &\sim \chi^2_{\operatorname{rank}\mathbb{H}}.\end{aligned}$$

### 6.1.1 *Independence of Least-Squares Coefficients and Residuals*

Assuming iid Normal errors, *the least-squares prediction vector* $\widehat{\mathbf{Y}}$ *is independent of the estimated error variance* $\hat{\sigma}^2$ defined in Sect. 4.2.1. The only randomness in $\hat{\sigma}^2$ comes from the residual vector $\mathbf{Y} - \widehat{\mathbf{Y}}$ which is the orthogonal projection of $\mathbb{E}\mathbf{Y} + \boldsymbol{\epsilon}$ onto $C(\mathbb{M})^{\perp}$. On the other hand, $\widehat{\mathbf{Y}}$ equals the orthogonal projection of $\mathbb{E}\mathbf{Y} + \boldsymbol{\epsilon}$ onto $C(\mathbb{M})$. The orthogonal projections of $\boldsymbol{\epsilon}$ onto orthogonal subspaces are independent (Sect. 5.1 and Fig. 6.1), so $\hat{\sigma}^2$ and $\widehat{\mathbf{Y}}$ must be independent. Importantly, the least-squares coefficients can also be written as a function of $\widehat{\mathbf{Y}}$ rather than $\mathbf{Y}$ (Exercise 2.8), so they are independent of $\hat{\sigma}^2$ as well.

### 6.1.2 *The Ratio Trick*

The fact that the error variance $\sigma^2$ is unknown seems to present a challenge for many statistical inference tasks. However, a wonderful trick is available to create random variables that do not depend on $\sigma^2$. I call it *the ratio trick*, because it involves devising a ratio of random variables for which the effects of $\sigma$ in the numerator and denominator cancel each other out.

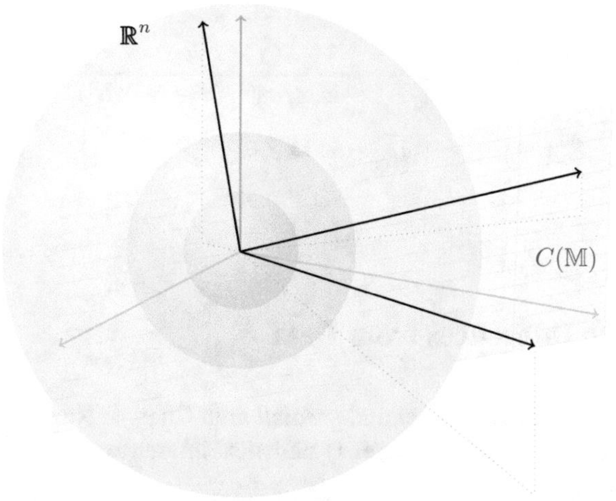


**Fig. 6.1** What if $\boldsymbol{\epsilon}$ is represented by another choice of orthogonal coordinate axes? By spherical symmetry, we can see that the joint distribution of the coordinate random variables must be the same regardless of the choice of orthogonal coordinate axes. In particular, one can consider a choice in which the first rank $\mathbb{M}$ axes are chosen from $C(\mathbb{M})$, and the remaining $n$ − rank $\mathbb{M}$ are (necessarily) chosen orthogonally to the $C(\mathbb{M})$. In the figure, the original axes have been grayed out, and new axes are drawn

**Theorem 6.1 (The Ratio Trick)** *Suppose* $\mathbf{Y} = \mathbb{M}\boldsymbol{\gamma} + \boldsymbol{\epsilon}$ *with* $\boldsymbol{\epsilon} \sim N(\mathbf{0}, \sigma^2\mathbb{I}_n)$*. Let* $\widehat{\mathbf{Y}}$ *denote the orthogonal projection of* $\mathbf{Y}$ *onto* $C(\mathbb{M})$*, and define* $\hat{\sigma}^2 := \|\mathbf{Y} - \widehat{\mathbf{Y}}\|^2/(n - \text{rank}\,\mathbb{M})$*. Then*

$$\frac{h(\widehat{\mathbf{Y}})}{\sigma} \sim N(0,1) \quad \Rightarrow \quad \frac{h(\widehat{\mathbf{Y}})}{\hat{\sigma}} \sim t_{n-\text{rank}\,\mathbb{M}}$$

*and*

$$\frac{h(\widehat{\mathbf{Y}})}{\sigma^2} \sim \chi^2_k \quad \Rightarrow \quad \frac{h(\widehat{\mathbf{Y}})/k}{\hat{\sigma}^2} \sim f_{k,n-\text{rank}\,\mathbb{M}}.$$

***Proof*** Let $\mathbb{H}$ denote the orthogonal projection matrix onto $C(\mathbb{M})$ and recall that $\mathbf{Y} - \widehat{\mathbf{Y}} = (\mathbb{I} - \mathbb{H})\mathbf{Y}$. Because $\mathbb{E}\mathbf{Y} \in C(\mathbb{M})$, we know that $(\mathbb{I} - \mathbb{H})\mathbf{Y} = (\mathbb{I} - \mathbb{H})\boldsymbol{\epsilon}$, which is independent of $h(\widehat{\mathbf{Y}})$ because they involve orthogonal projections of $\boldsymbol{\epsilon}$ onto orthogonal subspaces. The distribution of $\|(\mathbb{I} - \mathbb{H})\boldsymbol{\epsilon}/\sigma\|^2$ is $\chi^2_{n-\text{rank}\,\mathbb{M}}$.

If $h(\widehat{\mathbf{Y}})/\sigma \sim N(0,1)$, then

$$\frac{h(\widehat{\mathbf{Y}})}{\hat{\sigma}} = \frac{h(\widehat{\mathbf{Y}})/\sigma}{\|(\mathbb{I} - \mathbb{H})\boldsymbol{\epsilon}/\sigma\|/\sqrt{n - \text{rank}\,\mathbb{M}}} \sim t_{n-\text{rank}\,\mathbb{M}}.$$

Alternatively, if $\frac{h(\widehat{\mathbf{Y}})}{\sigma^2} \sim \chi^2_k$, then

$$\frac{h(\widehat{\mathbf{Y}})/k}{\hat{\sigma}^2} = \frac{h(\widehat{\mathbf{Y}})/\sigma^2 k}{\|(\mathbb{I} - \mathbb{H})\boldsymbol{\epsilon}/\sigma\|^2/(n - \text{rank}\,\mathbb{M})} \sim f_{k,n-\text{rank}\,\mathbb{M}}.$$

■

## 6.2 Making Inferences from Data

We have already done *estimation* and *prediction* in Chap. 4. Now we will move on to some additional tasks of (frequentist) statistical inference.

**Definition (Confidence Set)** Any subset of the parameter space that is a function of the explanatory and response variables and has a specified probability (often 0.95) of including the true parameter value is called a **confidence set**. The specified probability is called its **confidence level**.

A confidence set often takes the form of an interval and is then called a *confidence interval*. Sometimes the confidence *percent* is stated rather than the level; it is of course 100 times the confidence level.

**Definition (Test Statistic)** A **test statistic** is a function of the explanatory and response variables whose distribution can be derived (or approximated) based on a particular assumption (called the **null hypothesis**) about the distribution of the data. Ideally, the test statistic's distribution depends strongly on the veracity of the null hypothesis is true.

**Definition (Significance Probability)** The probability that a test statistic generated according to its null hypothesis would be at least as extreme as the observed value of the test statistic calculated from a particular dataset is called the test statistic's **significance probability** (for that dataset). Smaller significance probability indicates stronger evidence against the null hypothesis.

**Definition (Hypothesis Test)** A **hypothesis test** *rejects* the null hypothesis when the test statistic's significance probability is below a specified value (often 0.05) and *fails to reject* the null hypothesis otherwise. The specified probability is called its **significance level**.

We will need to make extensive use of cumulative distribution functions, so let us establish our notation in advance. We will let $\tau_k$ denote the cdf of the $t_k$-distribution and $\phi_{k,m}$ denote the cdf of the $f_{k,m}$-distribution.[1] The inverses of these function will also be crucial for devising confidence sets. To clarify the roles of these two functions, let $X \sim t_k$. Whereas $\tau_k(q)$ tells you the probability that $X \leq q$, $\tau_k^{-1}(p)$ tells you how large $q$ needs to be for the probability that $X \leq q$ to be $p$. This $q$ is called the *p-quantile* of the $t_k$ distribution.

### 6.2.1 Location

Suppose $Y_1, \ldots, Y_n \overset{iid}{\sim} N(\alpha, \sigma^2)$. The least-squares estimate for $\alpha$ is $\overline{Y} \sim N(\alpha, \sigma^2/n)$, and its standardized version is

$$\frac{\overline{Y} - \alpha}{\sigma/\sqrt{n}} \sim N(0, 1).$$

Because $\sigma$ is unknown, we cannot use this random variable for devising confidence intervals or for testing. However, because $\overline{Y}$ is a function of the orthogonal projection of $\boldsymbol{\epsilon}$ onto the span of $\mathbf{1}$, the *ratio trick* says that we can substitute $\hat{\sigma}$ for $\sigma$ to derive

[1] An extra parameter in the subscript will indicate non-centrality in Sect. 6.2.4.2.

$$\frac{\overline{Y} - \alpha}{\hat{\sigma}/\sqrt{n}} \sim t_{n-1}.$$

A confidence interval can be derived from this $t_{n-1}$-distributed random variable. For example, with $\tau_{n-1}^{-1}(0.975)$ representing the 0.975-quantile of the $t_{n-1}$ distribution,[2]

$$\mathbb{P}\left\{-\tau_{n-1}^{-1}(0.975) \leq \frac{\alpha - \overline{Y}}{\hat{\sigma}/\sqrt{n}} \leq \tau_{n-1}^{-1}(0.975)\right\} = 0.95,$$

where $\mathbb{P}$ maps events to their probabilities. The event can be rewritten as

$$\overline{Y} - \tau_{n-1}^{-1}(0.975)\frac{\hat{\sigma}}{\sqrt{n}} \leq \alpha \leq \overline{Y} + \tau_{n-1}^{-1}(0.975)\frac{\hat{\sigma}}{\sqrt{n}}$$

which means that $\overline{Y} \pm \tau_{n-1}^{-1}(0.975)\frac{\hat{\sigma}}{\sqrt{n}}$ is a 95% confidence interval for the true expectation $\alpha$.

Finally, consider the null hypothesis that $\alpha = 0$. This assumption leads us to the test statistic

$$\begin{aligned} T &:= \frac{\overline{Y}}{\hat{\sigma}/\sqrt{n}} \\ &\sim t_{n-1}. \end{aligned}$$

Its significance probability is $2\tau_{n-1}(-|T|)$, where $\tau_{n-1}$ is the cdf of the $t_{n-1}$ distribution. One-sided tests can similarly be devised, as can tests with any arbitrary hypothesized value for $\alpha$.

### 6.2.2 *Simple Linear Model Slope*

Let $\mathbf{x}$ be an explanatory variable with positive empirical variance $\sigma_{\mathbf{x}}^2$. Throughout this subsection, we will assume that the simple linear model holds

$$\mathbf{Y} = \alpha\mathbf{1} + \beta\mathbf{x} + \boldsymbol{\epsilon}$$

with iid $N(0, \sigma^2)$ errors. Let $\hat{\alpha}$ and $\hat{\beta}$ be the least-squares coefficients from simple linear regression.

[2] You will notice that this random variable has $\alpha - \overline{Y}$ in the numerator rather than $\overline{Y} - \alpha$. The negative of a $t$-distributed random variable has the same distribution due to the symmetry of $t$-distributions.

The distribution of the least-squares slope $\hat{\beta}$ is Normal with expectation $\beta$ and variance $\frac{\sigma^2}{n\sigma_{\mathbf{x}}^2}$ according to Sect. 4.1.2. The standardized version is

$$\frac{\hat{\beta}-\beta}{\sigma/\sqrt{n}\sigma_{\mathbf{x}}} \sim N(0,1).$$

The appearance of the unknown error variance prevents us from using this random variable for testing hypotheses or constructing confidence intervals. However, we can again invoke the ratio trick because $\hat{\beta}$ is a function of the orthogonal projection of $\boldsymbol{\epsilon}$ onto span$\{\mathbf{1},\mathbf{x}\}$ (from Exercise 2.8); if we substitute $\hat{\sigma}$ for $\sigma$,

$$\frac{\hat{\beta}-\beta}{\hat{\sigma}/\sqrt{n}\sigma_{\mathbf{x}}} \sim t_{n-2}$$

because span$\{\mathbf{1},\mathbf{x}\}$ has dimension 2.

We can use this $t_{n-2}$-distributed random variable to derive confidence intervals for the true coefficient's value (assuming the model is correctly specified). For example,

$$\mathbb{P}\left\{-\tau_{n-2}^{-1}(0.975) \le \frac{\beta-\hat{\beta}}{\frac{\hat{\sigma}}{\sqrt{n}\sigma_{\mathbf{x}}}} \le \tau_{n-2}^{-1}(0.975)\right\} = 0.95$$

The event can be rewritten as

$$\hat{\beta}-\tau_{n-2}^{-1}(0.975)\frac{\hat{\sigma}}{\sqrt{n}\sigma_{\mathbf{x}}} \le \beta \le \hat{\beta}+\tau_{n-2}^{-1}(0.975)\frac{\hat{\sigma}}{\sqrt{n}\sigma_{\mathbf{x}}}$$

which (recalling the formula for $\hat{\beta}$) means that $\rho_{\mathbf{x},\mathbf{Y}}\frac{\sigma_{\mathbf{Y}}}{\sigma_{\mathbf{x}}} \pm \tau_{n-2}^{-1}(0.975)\frac{\hat{\sigma}}{\sqrt{n}\sigma_{\mathbf{x}}}$ is a 95% confidence interval for the true slope $\beta$.

Finally, consider the null hypothesis that $\beta = 0$. This assumption leads us to the test statistic[3]

$$T_1 := \frac{\hat{\beta}}{\hat{\sigma}/\sqrt{n}\sigma_{\mathbf{x}}}$$
$$\sim t_{n-2}.$$

Its significance probability is $2\tau_{n-2}(-|T_1|)$.

[3]Substituting $\rho_{\mathbf{x},\mathbf{Y}}\frac{\sigma_{\mathbf{Y}}}{\sigma_{\mathbf{x}}}$ for $\hat{\beta}$, this statistic $T_1$ simplifies to $\frac{\rho_{\mathbf{x},\mathbf{Y}}\sigma_{\mathbf{Y}}}{\hat{\sigma}/\sqrt{n}}$.

### 6.2.3 Multiple Linear Model Slopes

Let the explanatory variables $\mathbf{x}^{(1)}, \ldots, \mathbf{x}^{(m)} \in \mathbb{R}^n$ be the columns of $\mathbb{X}$, let $\widetilde{\mathbb{X}}$ denote its centered version, and assume the empirical covariance matrix $\Sigma$ is full rank. Throughout this subsection, we will assume that the multiple linear model holds

$$\mathbf{Y} = \alpha\mathbf{1} + \widetilde{\mathbb{X}}\boldsymbol{\beta} + \boldsymbol{\epsilon}$$

with iid $N(0, \sigma^2)$ errors. Let $\hat{\alpha}$ and $\hat{\boldsymbol{\beta}}$ be the least-squares coefficients from multiple linear regression.

Practitioners are more interested in the coefficients of the explanatory variables than in the intercept, so while multiple linear regression does estimate an intercept, we will study inference tasks for the other coefficients. Recalling from Sect. 4.1.3 the expectation and covariance of the explanatory variables' least-squares coefficients, we conclude that with iid Normal errors, $\hat{\boldsymbol{\beta}} \sim N(\boldsymbol{\beta}, \frac{\sigma^2}{n}\Sigma^{-1})$.

#### 6.2.3.1 A Single Slope

Each least-squares slope $\hat{\beta}_1, \ldots, \hat{\beta}_m$ can be standardized analogously to the slope in the simple linear modeling context as described in Sect. 6.2.2.

**Exercise 6.3** Let $\widetilde{\mathbb{X}} \in \mathbb{R}^{n \times m}$ be a centered explanatory data matrix with full rank. Assume

$$\mathbf{Y} = \alpha + \widetilde{\mathbb{X}}\boldsymbol{\beta} + \boldsymbol{\epsilon}$$

with $\epsilon_1, \ldots, \epsilon_n \overset{iid}{\sim} N(0, \sigma^2)$. Write the standardized version of the least-squares slope $\hat{\beta}_j$. What is its distribution?

*Solution* The distribution of $\hat{\beta}_j$ is Normal because it is a linear transformation of $\mathbf{Y}$ which is Normal. Its expectation equals the $j$th entry of $\mathbb{E}\hat{\boldsymbol{\beta}} = \boldsymbol{\beta}$, and its variance equals the $j$th diagonal of $\operatorname{cov}\hat{\boldsymbol{\beta}} = \frac{\sigma^2}{n}\Sigma^{-1}$:

$$\hat{\beta}_j \sim N(\beta_j, \tfrac{\sigma^2}{n}\Sigma_{jj}^{-1}).$$

The standardized version is

$$\frac{\hat{\beta}_j - \beta_j}{\frac{\sigma}{\sqrt{n}}\sqrt{\Sigma_{jj}^{-1}}} \sim N(0, 1).$$

♦

**Exercise 6.4** Let $\widetilde{\mathbb{X}} \in \mathbb{R}^{n \times m}$ be a centered explanatory data matrix with full rank. Assume

$$\mathbf{Y} = \alpha + \widetilde{\mathbb{X}}\boldsymbol{\beta} + \boldsymbol{\epsilon}$$

with $\epsilon_1, \ldots, \epsilon_n \overset{iid}{\sim} N(0, \sigma^2)$. Devise a t-distributed random variable involving the least-squares slope $\hat{\beta}_j$.

*Solution* From Exercise 6.3, the standardized version is $\frac{\hat{\beta}_j - \beta_j}{\frac{\sigma}{\sqrt{n}}\sqrt{\Sigma^{-1}_{jj}}} \sim N(0, 1)$. Exercise 2.8 implies that $\hat{\boldsymbol{\beta}}$ is a function of $\mathbb{H}\boldsymbol{\epsilon}$, so the ratio trick allows us to substitute $\hat{\sigma}$ for $\sigma$ to derive

$$\frac{\hat{\beta}_j - \beta_j}{\frac{\hat{\sigma}}{\sqrt{n}}\sqrt{\Sigma^{-1}_{jj}}} \sim t_{n-m-1}.$$

♦

**Exercise 6.5** Let $\widetilde{\mathbb{X}} \in \mathbb{R}^{n \times m}$ be a centered explanatory data matrix with full rank. Assume

$$\mathbf{Y} = \alpha + \widetilde{\mathbb{X}}\boldsymbol{\beta} + \boldsymbol{\epsilon}$$

with $\epsilon_1, \ldots, \epsilon_n \overset{iid}{\sim} N(0, \sigma^2)$. Devise a 95% confidence interval for $\beta_j$.

*Solution* From Exercise 6.4, $\frac{\hat{\beta}_j - \beta_j}{\frac{\hat{\sigma}}{\sqrt{n}}\sqrt{\Sigma^{-1}_{jj}}} \sim t_{n-m-1}$, so

$$\mathbb{P}\left\{-\tau^{-1}_{n-m-1}(0.975) \leq \frac{\beta_j - \hat{\beta}_j}{\frac{\hat{\sigma}}{\sqrt{n}}\sqrt{\Sigma^{-1}_{jj}}} \leq \tau^{-1}_{n-m-1}(0.975)\right\} = 0.95.$$

The event can be rewritten as

$$\hat{\beta}_j - \tau^{-1}_{n-m-1}(0.975)\frac{\hat{\sigma}}{\sqrt{n}}\sqrt{\Sigma^{-1}_{jj}} \leq \beta_j \leq \hat{\beta}_j + \tau^{-1}_{n-m-1}(0.975)\frac{\hat{\sigma}}{\sqrt{n}}\sqrt{\Sigma^{-1}_{jj}}$$

which means that $\hat{\beta}_j \pm \tau^{-1}_{n-m-1}(0.975)\frac{\hat{\sigma}}{\sqrt{n}}\sqrt{\Sigma^{-1}_{jj}}$ is a 95% confidence interval for $\beta_j$. ♦

**Exercise 6.6** Let $\widetilde{\mathbb{X}} \in \mathbb{R}^{n \times m}$ be a centered explanatory data matrix with full rank. Assume

$$\mathbf{Y} = \alpha + \widetilde{\mathbb{X}}\boldsymbol{\beta} + \boldsymbol{\epsilon}$$

with $\epsilon_1, \ldots, \epsilon_n \overset{iid}{\sim} N(0, \sigma^2)$. Devise a test statistic $T_j$ for the null hypothesis that $\beta_j = 0$.

*Solution* From Exercise 6.4, $\frac{\hat{\beta}_j - \beta_j}{\frac{\hat{\sigma}}{\sqrt{n}}\sqrt{\Sigma^{-1}_{jj}}} \sim t_{n-m-1}$. The assumption that $\beta_j = 0$ leads to the test statistic

$$T_j := \frac{\hat{\beta}_j}{\frac{\hat{\sigma}}{\sqrt{n}}\sqrt{\Sigma^{-1}_{jj}}}$$

$$\sim t_{n-m-1}.$$

The significance probability is $2\tau_{n-m-1}(-|T_j|)$. ♦

#### 6.2.3.2 All Slopes Together

Next, let us consider *all* of the explanatory variables' coefficients together. We will parallel the logic of the previous section by first standardizing the random vector:

$$\frac{\sqrt{n}}{\sigma}\Sigma^{1/2}\left(\hat{\boldsymbol{\beta}} - \boldsymbol{\beta}\right) \sim N(\mathbf{0}, \mathbb{I}_m).$$

The squared norm of this random vector is $\chi^2_m$-distributed (Sect. 5.2.2). As before, the unknown error variance prevents us from making direct use of these quantities for inference. Also as before, the ratio trick lets us substitute $\hat{\sigma}$ for $\sigma$ to derive

$$\frac{n\left\|\Sigma^{1/2}\left(\hat{\boldsymbol{\beta}} - \boldsymbol{\beta}\right)\right\|^2/m}{\hat{\sigma}^2} \sim f_{m,n-m-1}.$$

We can use our $f_{m,n-m-1}$-distributed random variable[4] to derive confidence sets for the true coefficient vector. For example, with $\phi^{-1}_{m,n-m-1}(0.95)$ representing the 0.95-quantile of the $f_{m,n-m-1}$ distribution,

$$\mathbb{P}\left\{\frac{n\|\Sigma^{1/2}(\boldsymbol{\beta} - \hat{\boldsymbol{\beta}})\|^2/m}{\hat{\sigma}^2} \le \phi^{-1}_{m,n-m-1}(0.95)\right\} = 0.95.$$

The event can be rewritten as

$$\left\|\frac{\sqrt{n}}{\hat{\sigma}}\Sigma^{1/2}(\boldsymbol{\beta} - \hat{\boldsymbol{\beta}})\right\|^2 \le m\phi^{-1}_{m,n-m-1}(0.95).$$

[4]The confidence set derivation actually replaces the original random vector with its negative which has the exact same squared length.

Thus, we have a 95% confidence ellipsoid for $\boldsymbol{\beta}$ comprising the set of vectors whose squared Mahalanobis distance from $N(\hat{\boldsymbol{\beta}}, \frac{\hat{\sigma}^2}{n}\Sigma^{-1})$ is no greater than $m\phi^{-1}_{m,n-m-1}(0.95)$. (Notice that $N(\hat{\boldsymbol{\beta}}, \frac{\hat{\sigma}^2}{n}\Sigma^{-1})$ is a sensible estimate for the distribution of $\hat{\boldsymbol{\beta}}$; it substitutes our estimates for the true distribution's unknown parameters.)

There is a dual relationship between the hypothesis tests and confidence ellipsoids here. A hypothesized coefficient vector $\boldsymbol{\beta}_{\text{null}}$ will be rejected by the 0.05-level test if and only if it lies outside of the 95% confidence ellipsoid.

Let us also consider the common null hypothesis that $\boldsymbol{\beta} = \mathbf{0}$. If this were true, then the test statistic

$$
\begin{aligned}
F &:= \frac{n\|\Sigma^{1/2}\hat{\boldsymbol{\beta}}\|^2/m}{\hat{\sigma}^2} \\
&\sim f_{m,n-m-1}
\end{aligned}
\tag{6.1}
$$

can be used. The significance probability is $1 - \phi_{m,n-m-1}(F)$, where $\phi_{m,n-m-1}$ is the cdf of the $f_{m,n-m-1}$ distribution.[5]

#### 6.2.3.3 Prediction Intervals

**Definition (Prediction Set)** Let $X$ be a random element taking values in a set $S$. Any subset of $S$ that is a function of the explanatory and response variables and has a specified probability (often 0.95) according to the distribution of $X$ is called a **prediction set**. The specified probability is called its **prediction level**.

Let us revisit the prediction introduced in Sects. 4.1.4.2 and 4.2.2. A new response $Y_{n+1}$ is to be predicted by $\widehat{Y}_{n+1}$ which uses the multiple linear regression coefficients of the previous $n$ observations along with the explanatory values of the new observation. We found that $Y_{n+1} - \widehat{Y}_{n+1}$ has expectation 0 and variance $\sigma^2[1 + \frac{1}{n}(1 + \delta^2)]$ where $\delta$ is the Mahalanobis distance from the new observation's vector of explanatory values to the empirical distribution of the original explanatory data. In fact, if $\epsilon_1, \ldots, \epsilon_{n+1} \overset{iid}{\sim} N(0, \sigma^2)$, then $Y_{n+1} - \widehat{Y}_{n+1}$ is a difference of independent Normal random variables, so we know more specifically that its distribution is $N(0, \sigma^2[1 + \frac{1}{n}(1 + \delta^2)])$. Using steps analogous to our earlier confidence interval derivations, we can actually formulate *prediction intervals* for $\widehat{Y}_{n+1}$.

[5]This single test of the coefficients simultaneously is not the same as testing all of the coefficients separately. Performing multiple different tests results in an *overall* false positive probability that is larger than the false positive rates of the individual tests as will be discussed in Sect. 6.2.4.

The standardized version of prediction error is

$$\frac{Y_{n+1} - \widehat{Y}_{n+1}}{\sigma\sqrt{1 + \frac{1}{n}(1+\delta^2)}} \sim N(0, 1).$$

Because $\hat{\sigma}^2$ is independent of $\epsilon_{n+1}$ and of the least-squares coefficients, we can follow the same pattern as before, substituting $\hat{\sigma}$ for $\sigma$.

$$\frac{Y_{n+1} - \widehat{Y}_{n+1}}{\hat{\sigma}\sqrt{1 + \frac{1}{n}(1+\delta^2)}} \sim t_{n-m-1}.$$

Therefore

$$\mathbb{P}\left\{-\tau_{n-m-1}^{-1}(0.975) \leq \frac{Y_{n+1} - \widehat{Y}_{n+1}}{\hat{\sigma}\sqrt{1 + \frac{1}{n}(1+\delta^2)}} \leq \tau_{n-m-1}^{-1}(0.975)\right\} = 0.95.$$

The event can be rewritten as

$$\widehat{Y}_{n+1} - \tau_{n-m-1}^{-1}(0.975)\hat{\sigma}\sqrt{1 + \tfrac{1}{n}(1+\delta^2)} \leq Y_{n+1}$$
$$\leq \widehat{Y}_{n+1} + \tau_{n-m-1}^{-1}(0.975)\hat{\sigma}\sqrt{1 + \tfrac{1}{n}(1+\delta^2)}$$

meaning that $\widehat{Y}_{n+1} \pm \tau_{n-m-1}^{-1}(0.975)\hat{\sigma}\sqrt{1 + \frac{1}{n}(1+\delta^2)}$ is a 95% prediction interval for $Y_{n+1}$.

Let us contrast this with a *confidence interval* for the *expectation* of the new response value (Fig. 6.2), which can be derived similarly. We can see that $\hat{Y}_{n+1}$ is Normal, and its expectation and variance were found in Sect. 4.2.2. Standardizing, we have

$$\frac{\hat{Y}_{n+1} - \mathbb{E}Y_{n+1}}{\sigma\sqrt{\frac{1}{n}(1+\delta^2)}} \sim N(0, 1).$$

Because $\hat{\sigma}^2$ is independent of this random variable,

$$\frac{\hat{Y}_{n+1} - \mathbb{E}Y_{n+1}}{\hat{\sigma}\sqrt{\frac{1}{n}(1+\delta^2)}} = \frac{(\hat{Y}_{n+1} - \mathbb{E}Y_{n+1})/(\sigma\sqrt{\frac{1}{n}(1+\delta^2)})}{\hat{\sigma}/\sigma}$$
$$\sim t_{n-m-1}.$$

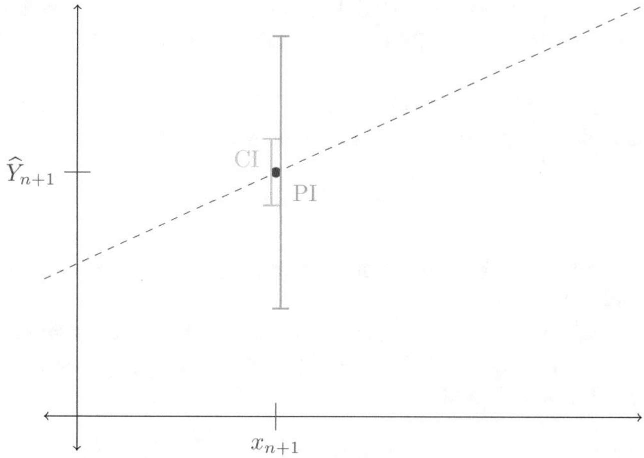


**Fig. 6.2** A comparison of a confidence interval (CI) and a prediction interval (PI) is drawn in the context of simple linear modeling. The confidence interval for $\mathbb{E}\mathbf{Y}_{n+1}$ and prediction interval for $Y_{n+1}$ are both centered at the least-squares line's value at the new explanatory value $x_{n+1}$. The prediction interval necessarily extends further, and its width does not decrease to zero

Therefore

$$\mathbb{P}\left\{-\tau_{n-m-1}^{-1}(0.975) \le \frac{\mathbb{E}Y_{n+1} - \hat{Y}_{n+1}}{\hat{\sigma}\sqrt{\frac{1}{n}(1+\delta^2)}} \le \tau_{n-m-1}^{-1}(0.975)\right\} = 0.95.$$

The event can be rewritten as

$$\hat{Y}_{n+1} - \tau_{n-m-1}^{-1}(0.975)\hat{\sigma}\sqrt{\tfrac{1}{n}(1+\delta^2)} \le \mathbb{E}Y_{n+1} \le \hat{Y}_{n+1} + \tau_{n-m-1}^{-1}(0.975)\hat{\sigma}\sqrt{\tfrac{1}{n}(1+\delta^2)}$$

which shows that $\hat{Y}_{n+1} \pm \tau_{n-m-1}^{-1}(0.975)\hat{\sigma}\sqrt{\frac{1}{n}(1+\delta^2)}$ constitutes a 95% confidence interval for $\mathbb{E}Y_{n+1}$.

## 6.2.4 *Linear Models and Submodels*

The hypothesis that some coefficient equals 0 is identical to the hypothesis that $\mathbb{E}\mathbf{Y}$ lies in the span of the other columns of the design matrix. Similarly, the hypothesis that $\beta_1, \ldots, \beta_m$ are all 0 in a multiple linear model is identical to the hypothesis that $\mathbb{E}\mathbf{Y}$ lies in the span of $\mathbf{1}$. While these two hypotheses can easily be incorporated into our earlier derivations from Sect. 6.2.3.1 and 6.2.3.2, there is a general formula for testing the hypothesis that $\mathbb{E}\mathbf{Y}$ lies in any particular subspace of $C(\mathbb{M})$ (sometimes called a *submodel*).

In this section we will use the more general formulation of the linear model. For some design matrix $\mathbb{M}$, the expectation of $\mathbf{Y}$ is in $C(\mathbb{M})$ and[6]

$$\mathbf{Y} = \mathbb{E}\mathbf{Y} + \boldsymbol{\epsilon}$$

with $\boldsymbol{\epsilon} \sim N(\mathbf{0}, \sigma^2\mathbb{I})$. Let $\widehat{\mathbf{Y}}$ be the least-squares prediction vector when the model is correctly specified.

Let $\mathcal{S}$ be a subspace of $C(\mathbb{M})$, and let $\mathbb{H}$ and $\mathbb{H}_{\mathcal{S}}$ be the orthogonal projection matrices onto $C(\mathbb{M})$ and $\mathcal{S}$. Let $\check{\mathbf{Y}}$ denote the orthogonal projection of $\mathbf{Y}$ onto $\mathcal{S}$. The difference between $\widehat{\mathbf{Y}}$ and $\check{\mathbf{Y}}$ is

$$\begin{aligned} \widehat{\mathbf{Y}} - \check{\mathbf{Y}} &= \mathbb{H}\mathbf{Y} - \mathbb{H}_{\mathcal{S}}\mathbf{Y} \\ &= (\mathbb{H} - \mathbb{H}_{\mathcal{S}})\mathbf{Y} \\ &= (\mathbb{H} - \mathbb{H}_{\mathcal{S}})(\mathbb{E}\mathbf{Y} + \boldsymbol{\epsilon}) \\ &= (\mathbb{H} - \mathbb{H}_{\mathcal{S}})\mathbb{E}\mathbf{Y} + (\mathbb{H} - \mathbb{H}_{\mathcal{S}})\boldsymbol{\epsilon}. \end{aligned} \tag{6.2}$$

From Exercise 1.49, $\mathbb{H} - \mathbb{H}_{\mathcal{S}}$ is the orthogonal projection matrix onto the orthogonal complement of $\mathcal{S}$ within $C(\mathbb{M})$. This subspace is in $C(\mathbb{M})$, so with iid Normal errors, we know that $(\mathbb{H} - \mathbb{H}_{\mathcal{S}})\boldsymbol{\epsilon}$ must be independent of $\hat{\sigma}^2$.

Now assume that $\mathbb{E}\mathbf{Y} \in \mathcal{S}$. Then the orthogonal projection of $\mathbb{E}\mathbf{Y}$ onto the orthogonal complement of $\mathcal{S}$ within $C(\mathbb{M})$ is $(\mathbb{H} - \mathbb{H}_{\mathcal{S}})\mathbb{E}\mathbf{Y} = \mathbf{0}$, so

$$\begin{aligned} \frac{\|\widehat{\mathbf{Y}} - \check{\mathbf{Y}}\|^2/(\operatorname{rank}\mathbb{M} - \dim\mathcal{S})}{\hat{\sigma}^2} &= \frac{\|(\mathbb{H} - \mathbb{H}_{\mathcal{S}})\boldsymbol{\epsilon}\|^2/(\operatorname{rank}\mathbb{M} - \dim\mathcal{S})}{\|(\mathbb{I} - \mathbb{H})\boldsymbol{\epsilon}\|^2/(n - \operatorname{rank}\mathbb{M})} \\ &\sim f_{\operatorname{rank}\mathbb{M} - \dim\mathcal{S},\, n - \operatorname{rank}\mathbb{M}} \end{aligned} \tag{6.3}$$

according to Exercise 5.12. Notice that such a test does not require the design matrix to be full rank or to include span$\{\mathbf{1}\}$ in its column space.

**Exercise 6.7** Section 6.2.3.2 described a test statistic (Eq. 6.1) for the null hypothesis that all of the slopes in a multiple linear model are 0. Is it the same as the test statistic prescribed by Eq. 6.3?

*Solution* The null hypothesis is that $\mathbb{E}\mathbf{Y}$ is in the span of $\mathbf{1}$, so the general approach (Eq. 6.3) uses the test statistic

[6] I have avoided assigning a symbol to the coefficient vector to emphasize that they will not play any role in this testing procedure.

$$\frac{\|\widehat{\mathbf{Y}} - \overline{Y}\mathbf{1}\|^2/m}{\hat{\sigma}^2} \sim f_{m,n-m-1},$$

while Sect. 6.2.3.2 derived the test statistic

$$\frac{n\|\Sigma^{1/2}\hat{\boldsymbol{\beta}}\|^2/m}{\hat{\sigma}^2} \sim f_{m,n-m-1}.$$

Let us analyze the factor in which they appear to differ. Recall that for multiple linear regression the least-squares prediction vector can be expressed as

$$\widehat{\mathbf{Y}} = \overline{Y}\mathbf{1} + \widetilde{\mathbb{X}}\hat{\boldsymbol{\beta}}$$

where $\widetilde{\mathbb{X}}$ is the centered explanatory data matrix. Therefore,

$$\begin{aligned}\|\widehat{\mathbf{Y}} - \overline{Y}\mathbf{1}\|^2 &= \|\widetilde{\mathbb{X}}\hat{\boldsymbol{\beta}}\|^2 \\ &= \hat{\boldsymbol{\beta}}^T \widetilde{\mathbb{X}}^T \widetilde{\mathbb{X}} \hat{\boldsymbol{\beta}}.\end{aligned}$$

And in the other test statistic,

$$\begin{aligned}n\|\Sigma^{1/2}\hat{\boldsymbol{\beta}}\|^2 &= n\hat{\boldsymbol{\beta}}^T \underbrace{\Sigma}_{\frac{1}{n}\widetilde{\mathbb{X}}^T\widetilde{\mathbb{X}}} \hat{\boldsymbol{\beta}} \\ &= \hat{\boldsymbol{\beta}}^T \widetilde{\mathbb{X}}^T \widetilde{\mathbb{X}} \hat{\boldsymbol{\beta}}\end{aligned}$$

so they turn out to be exactly the same. ♦

#### 6.2.4.1 Testing with Categorical Explanatory Variables

**Theorem 6.2 (One-way ANOVA Test)** *Let $Y_1, \ldots, Y_n$ be independent draws from k groups such that $Y_i \sim N(\alpha_j, \sigma^2)$ if $Y_i$ belongs to group j. Let $n_1, \ldots, n_k$ be the number of observations in each group. Let $\overline{Y}$ denote the overall average and $\overline{Y}_1, \ldots, \overline{Y}_k$ denote the groups' averages. Under the null hypothesis that all the groups have the same expectation,*

$$\frac{SS_{reg}/(k-1)}{SS_{res}/(n-k)} \sim f_{k-1,n-k},$$

*where $SS_{reg}$ is the* regression sum of squares

$$SS_{reg} = \sum_{j=1}^{k} n_j(\overline{Y}_j - \overline{Y})^2$$

*and $SS_{res}$ is the* residual sum of squares

$$SS_{res} = \sum_{j=1}^{k} \sum_{i:x_i=j} (Y_i - \overline{Y}_j)^2.$$

***Proof*** Recall from Chaps. 2 and 4 how to represent this scenario with a linear model. If the $i$th observation belongs to group $j$, then the $i$th row of the design matrix $\mathbb{M}$ has a 1 in its column $j$ entry and zeros elsewhere. The design matrix has $k$ linearly independent columns and those columns sum to $\mathbf{1}$, so $\mathbf{1} \in C(\mathbb{M})$. The least-squares procedure predicts each response by its group's average.

As in Sect. 2.2.1, the sums of squares can also be represented as squared norms of vectors. Summing over observations rather than groups (exactly as in Eq. 2.1), the regression sum of squares is

$$\begin{aligned}\sum_{j=1}^{k} n_j(\overline{Y}_j - \overline{Y})^2 &= \sum_i (\overline{Y}_{x_i} - \overline{Y})^2 \\ &= \sum_i (\widehat{Y}_i - \overline{Y})^2 \\ &= \|\widehat{\mathbf{Y}} - \overline{Y}\mathbf{1}\|^2.\end{aligned}$$

Similarly, the residual sum of squares is

$$\begin{aligned}\sum_{j=1}^{k} \sum_{i:x_i=j} (Y_i - \overline{Y}_j)^2 &= \sum_i (Y_i - \overline{Y}_{x_i})^2 \\ &= \sum_i (Y_i - \widehat{Y}_i)^2 \\ &= \|\mathbf{Y} - \widehat{\mathbf{Y}}\|^2.\end{aligned}$$

At this point, we can recognize squared norm of the least-squares residual vector divided by $n - k$ as $\hat{\sigma}^2$, so we can express the test statistic as

$$\frac{\|\widehat{\mathbf{Y}} - \overline{Y}\mathbf{1}\|^2/(k-1)}{\hat{\sigma}^2}.$$

The null hypothesis is that the groups all share the same expectation $\alpha_1 = \ldots = \alpha_k$. That is equivalent to every entry of $\mathbb{E}\mathbf{Y}$ being the same number, i.e. $\mathbb{E}\mathbf{Y}$ being in the span of $\{\mathbf{1}\}$. The projection of $\mathbf{Y}$ onto that subspace is $\overline{Y}\mathbf{1}$, so the one-way ANOVA test is an instance of Eq. 6.3. ■

When the ANOVA test rejects its null hypothesis, you conclude that the groups *do not* all share the same expectation. However, simply establishing that the groups are different is not necessarily very useful. You will likely want to make more specific comparisons among the groups. It is easy to test for equality of expectation for every pair of groups. This would produce $\binom{k}{2} = \frac{k^2-k}{2}$ significance probabilities, which grows quadratically with the number of groups. Whenever *multiple different tests* are performed, the *overall* false positive rate is larger than the false positive rate of any of the individual tests. To keep the overall false positive rate under control and limit the likely number of false positives, practitioners often use a combination of sequential testing (i.e. some of the tests are only performed if a preceding test rejects its null hypothesis), limiting the total number of tests, and using smaller false positive probabilities for the individual tests.

To this end, innumerable such schemes have been suggested. The most conservative approach is the *Bonferroni correction* method: the sum of the individual tests' false positive probabilities should equal the overall false positive rate that you are willing to tolerate. For example, if you perform $m$ tests, each with false positive probability $p/m$, then the overall false positive rate is no greater than $p$ (Exercise 3.1). On the other hand, if the $m$ test statistics are *independent* of each other, then *Sidak's correction* can be applied: a false positive rate of $1-(1-p)^{1/m}$ is used for each of the tests, resulting in an overall false positive rate of exactly $p$ (substitute this into Exercise 3.13 to check).

With one categorical explanatory variable and balanced groups, a preferred method for multiple testing is to use *orthogonal contrasts*. Let $\mathbf{w} := (w_1, \ldots, w_k) \in \mathbb{R}^k$, and let the design matrix $\mathbb{M}$ be defined as in Sect. 2.2.2. Consider the null hypothesis that $\sum_{j=1}^k w_j \alpha_j = 0$, i.e. $\boldsymbol{\alpha} := (\alpha_1, \ldots, \alpha_k) \perp \mathbf{w}$. This statement about $\boldsymbol{\alpha} \in \mathbb{R}^k$ is equivalent to a statement about $\mathbb{E}\mathbf{Y} \in \mathbb{R}^n$, namely, that $\mathbb{E}\mathbf{Y} \perp \mathbb{M}\mathbf{w}$. This is because the inner product of $\mathbb{E}\mathbf{Y}$ with $\mathbb{M}\mathbf{w}$ is proportional to the inner product of $\boldsymbol{\alpha}$ with $\mathbf{w}$:

$$\begin{aligned}\langle \mathbb{E}\mathbf{Y}, \mathbb{M}\mathbf{w}\rangle &= \sum_{i=1}^n w_{x_i}\alpha_{x_i} \\ &= (n/k)\underbrace{\sum_{j=1}^k w_j\alpha_j}_{\langle \boldsymbol{\alpha}, \mathbf{w}\rangle}.\end{aligned}$$

As a specific example, suppose there are three groups, and consider the two orthogonal contrasts $\mathbf{w}^{(1)} := (1, -1, 0)$ and $\mathbf{w}^{(2)} := (\frac{1}{2}, \frac{1}{2}, -1)$. The hypothesis $\mathbb{E}\mathbf{Y} \perp \mathbb{M}\mathbf{w}^{(1)}$ is equivalent to $\alpha_1 = \alpha_2$, while the hypothesis $\mathbb{E}\mathbf{Y} \perp \mathbb{M}\mathbf{w}^{(2)}$ is equivalent to $\frac{1}{2}\alpha_1 + \frac{1}{2}\alpha_2 = \alpha_3$.

Next, let us derive a test statistic for any such *contrast hypothesis*. The coordinate of $\mathbf{Y}$ in the direction of $\mathbb{M}\mathbf{w}$ is

$$\left\langle \mathbf{Y}, \frac{\mathbb{M}\mathbf{w}}{\|\mathbb{M}\mathbf{w}\|} \right\rangle = \left\langle \mathbb{E}\mathbf{Y} + \boldsymbol{\epsilon}, \frac{\mathbb{M}\mathbf{w}}{\|\mathbb{M}\mathbf{w}\|} \right\rangle$$
$$= \left\langle \mathbb{E}\mathbf{Y}, \frac{\mathbb{M}\mathbf{w}}{\|\mathbb{M}\mathbf{w}\|} \right\rangle + \left\langle \boldsymbol{\epsilon}, \frac{\mathbb{M}\mathbf{w}}{\|\mathbb{M}\mathbf{w}\|} \right\rangle.$$

Furthermore, because $\mathbb{M}\mathbf{w} \in C(\mathbb{M})$, this coordinate is independent of $\hat{\sigma}^2$. If $\mathbb{E}\mathbf{Y} \perp \mathbb{M}\mathbf{w}$, then this coordinate simplifies to $\langle \boldsymbol{\epsilon}, \frac{\mathbb{M}\mathbf{w}}{\|\mathbb{M}\mathbf{w}\|} \rangle$, the coordinate of the error vector in the direction of $\mathbb{M}\mathbf{w}$. With $\boldsymbol{\epsilon} \sim N(\mathbf{0}, \sigma^2\mathbb{I})$, this coordinate divided by $\hat{\sigma}$ is

$$\frac{\langle \boldsymbol{\epsilon}, \frac{\mathbb{M}\mathbf{w}}{\|\mathbb{M}\mathbf{w}\|} \rangle}{\hat{\sigma}} = \frac{\langle \boldsymbol{\epsilon}, \frac{\mathbb{M}\mathbf{w}}{\|\mathbb{M}\mathbf{w}\|} \rangle}{\|(\mathbb{I} - \mathbb{H})\boldsymbol{\epsilon}\| / \sqrt{n-k}}$$
$$\sim t_{n-k}$$

by Exercise 5.9. Using either Exercise 5.11 or Exercise 5.12, its square is seen to be

$$\frac{\langle \boldsymbol{\epsilon}, \frac{\mathbb{M}\mathbf{w}}{\|\mathbb{M}\mathbf{w}\|} \rangle^2}{\hat{\sigma}^2} = \frac{\langle \boldsymbol{\epsilon}, \frac{\mathbb{M}\mathbf{w}}{\|\mathbb{M}\mathbf{w}\|} \rangle^2 / 1}{\|(\mathbb{I} - \mathbb{H})\boldsymbol{\epsilon}\|^2 / (n-k)}$$
$$\sim f_{1,n-k}.$$

With a simplified expression for $\langle \mathbf{Y}, \frac{\mathbb{M}\mathbf{w}}{\|\mathbb{M}\mathbf{w}\|} \rangle^2$ (see Eq. 2.2), we have the following test statistic for the null hypothesis that $\mathbb{E}\mathbf{Y} \perp \mathbb{M}\mathbf{w}$, i.e. that $\sum_{j=1}^k w_j \alpha_j = 0$:

$$\frac{(n/k)[\sum_{j=1}^k w_j \overline{Y}_j]^2 / \sum_{j=1}^k w_j^2}{\hat{\sigma}^2} \sim f_{1,n-k}.$$

With $k - 1$ orthogonal contrasts, we have seen that the squared contrast coefficients add up to the regression sum of squares. Thus the ANOVA test statistic is the average of the contrast test statistics. Additionally, because they involve projections of $\boldsymbol{\epsilon}$ onto orthogonal subspaces, the numerators of the contrast test statistics are independent of each other. The contrast test statistics are not entirely independent because of their shared dependence on $\hat{\sigma}^2$, but they can be considered nearly independent especially if the sample size is large. As a result, statisticians commonly use Sidak's correction when testing *planned* orthogonal contrasts.[7]

Orthogonal contrast testing can be difficult to understand in the abstract; it will make more sense if you work through an example.[8]

[7]If the contrasts were designed after the response values were seen (called *post hoc* testing), then Sidak's correction does not produce the specified overall false positive rate. In that case, sequential testing (after the ANOVA test) is a reasonable alternative approach.

[8]Homework tasks involving contrast testing are available on this book's website.

#### 6.2.4.2 Power

**Definition (Power)** The **power** of a hypothesis test for a particular alternative hypothesis is the probability that the test will reject the null hypothesis when that alternative is true.

The power can be calculated if one can determine the distribution of the test statistic when the alternative hypothesis is true. For linear modeling, that generally means making use of the *non-central* distributions defined in Chap. 5. Let us see how this works for the general subspace test described earlier in Sect. 6.2.4. With $\widehat{\mathbf{Y}}$ representing the orthogonal projection of $\mathbf{Y}$ onto the column space of the design matrix $\mathbb{M}$ and with $\check{\mathbf{Y}}$ representing the orthogonal projection of $\mathbf{Y}$ onto a smaller subspace $\mathcal{S} \subseteq C(\mathbb{M})$, we can represent their difference as

$$\begin{aligned}\widehat{\mathbf{Y}} - \check{\mathbf{Y}} &= (\mathbb{H} - \mathbb{H}_0)\mathbf{Y}\\ &= (\mathbb{H} - \mathbb{H}_0)(\mathbb{E}\mathbf{Y} + \boldsymbol{\epsilon})\\ &= (\mathbb{H} - \mathbb{H}_0)\mathbb{E}\mathbf{Y} + (\mathbb{H} - \mathbb{H}_0)\boldsymbol{\epsilon}.\end{aligned}$$

If $\boldsymbol{\epsilon} \sim N(\mathbf{0}, \sigma^2\mathbb{I})$, then we can make use of Exercise 5.13 to conclude that the $f$-statistic is

$$\begin{aligned}\frac{\|\widehat{\mathbf{Y}} - \check{\mathbf{Y}}\|^2/(\operatorname{rank}\mathbb{M} - \dim\mathcal{S})}{\hat{\sigma}^2} &= \frac{\|(\mathbb{H} - \mathbb{H}_0)\mathbb{E}\mathbf{Y} + (\mathbb{H} - \mathbb{H}_0)\boldsymbol{\epsilon}\|^2/(\operatorname{rank}\mathbb{M} - \dim\mathcal{S})}{\hat{\sigma}^2}\\ &\sim f_{\operatorname{rank}\mathbb{M}-\dim\mathcal{S},\, n-\operatorname{rank}\mathbb{M},\, \|(\mathbb{H}-\mathbb{H}_0)\mathbb{E}\mathbf{Y}\|/\sigma^2}.\end{aligned}$$

To test with false positive rate 0.05, one would reject the null hypothesis precisely when the $f$-statistic is greater than $\phi^{-1}_{\operatorname{rank}\mathbb{M}-\dim\mathcal{S},\, n-\operatorname{rank}\mathbb{M}}(0.95)$, the 0.95 quantile of

$$f_{\operatorname{rank}\mathbb{M}-\dim\mathcal{S},\, n-\operatorname{rank}\mathbb{M}}.$$

Therefore, the power of the test for an alternative specification of $\mathbb{E}\mathbf{Y}$ and $\sigma^2$ is

$$1 - \phi_{\operatorname{rank}\mathbb{M}-\dim\mathcal{S},\, n-\operatorname{rank}\mathbb{M},\, \|(\mathbb{H}-\mathbb{H}_0)\mathbb{E}\mathbf{Y}\|/\sigma^2}(\phi^{-1}_{\operatorname{rank}\mathbb{M}-\dim\mathcal{S},\, n-\operatorname{rank}\mathbb{M}}(0.95)),$$

where $\phi_{\operatorname{rank}\mathbb{M}-\dim\mathcal{S},\, n-\operatorname{rank}\mathbb{M},\, \|(\mathbb{H}-\mathbb{H}_0)\mathbb{E}\mathbf{Y}\|/\sigma^2}$ is the cdf of

$$f_{\operatorname{rank}\mathbb{M}-\dim\mathcal{S},\, n-\operatorname{rank}\mathbb{M},\, \|(\mathbb{H}-\mathbb{H}_0)\mathbb{E}\mathbf{Y}\|/\sigma^2}.$$

While we did not need the null hypothesis to specify the error variance $\sigma^2$, it does need to be specified by an alternative hypothesis for the power calculation.

**Exercise 6.8** Let $\mathbf{Y} = \alpha\mathbf{1} + \widetilde{\mathbb{X}}\boldsymbol{\beta} + \boldsymbol{\epsilon}$ with $\widetilde{\mathbb{X}} \in \mathbb{R}^{n \times m}$ representing a centered data matrix. If $\widehat{\mathbf{Y}}$ is the least-squares prediction vector that comes from multiple linear regression, find the distribution of

$$\frac{\|\widehat{\mathbf{Y}} - \overline{Y}\mathbf{1}\|^2}{\hat{\sigma}^2}.$$

*Solution* Based on the preceding discussion, the statistic in question has non-central $f$-distribution. $\widehat{\mathbf{Y}}$ is the projection onto an $(m + 1)$-dimensional subspace, while $\overline{Y}\mathbf{1}$ is the projection onto a 1-dimensional subspace. Thus the numerator has $m$ degrees of freedom, and the denominator has $n - m - 1$ degrees of freedom. The non-centrality parameter is

$$\|(\alpha\mathbf{1} + \widetilde{\mathbb{X}}\boldsymbol{\beta}) - (\alpha\mathbf{1})\|^2/\sigma^2 = \|\widetilde{\mathbb{X}}\boldsymbol{\beta}\|^2/\sigma^2.$$

♦

# Appendix A
# Epilogue

This completes our coverage of the core theory of linear models. The focus was on developing the reader's ability to visualize data in two important ways: as *observations* and as *variables*. The *observations* picture is more natural and intuitive, while understanding the *variables* picture involves something of a mental breakthrough. Both approaches are tremendously valuable in understanding linear model theory.

Remarkably, random variables can also be profitably understood with two pictures that are perfectly analogous to those we have been studying. A random variable defined on a probability space has a distribution on $\mathbb{R}$ (the observations picture), but it can also be thought of as a vector in the space of all possible random variables on that probability space (the variables picture). Understanding this can revolutionize the way you think about probability theory. Additionally, it generalizes many of the results that we have developed in this book; these results are easily seen as special cases. So if you are bold enough to pursue the next mental breakthrough, continue your journey with *Visualizing Random Vectors* (currently still in preparation, as of this book's printing).


W. D. Brinda, *Visualizing Linear Models*,
https://doi.org/10.1007/978-3-030-64167-2

# Appendix B
# Key Figures

Some of the most important figures from the text are recollected here for the reader's convenience. Study these tables of pictures until you comprehend clearly the distinction between the *observations picture* and the *variables picture*. Looking at these figures together might also better help you understand the progression from data-only regression (Chap. 2) to modeling (Chap. 4) as well as the progression from location regression/model to simple linear regression/model to multiple linear regression/model.


W. D. Brinda, *Visualizing Linear Models*,
https://doi.org/10.1007/978-3-030-64167-2

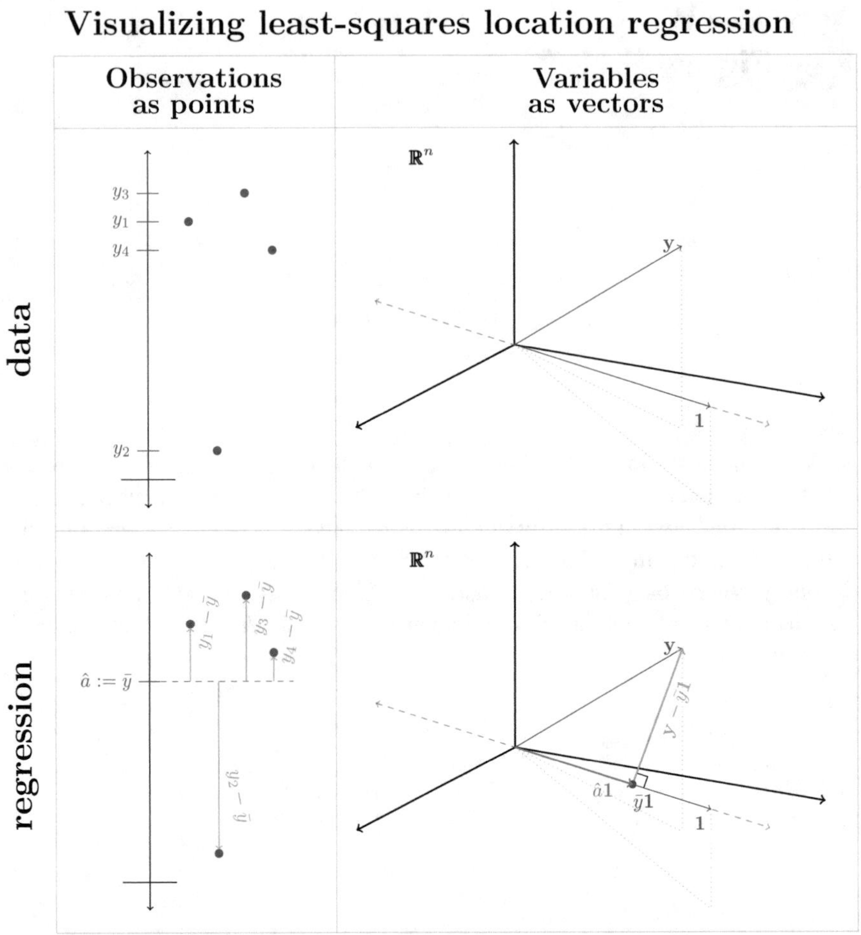


**Fig. 2.1 Observations as points.** *data*: A one-dimensional scatterplot draws points with heights equal to the response values. (The points are spaced out horizontally only to help us see them better.) *regression*: Arrows represent residuals when the response values are all predicted by the height of the dotted line. The average of the response values is the least-squares point, the prediction that produces the smallest sum of squared residuals. **Variables as vectors.** *data*: A three-dimensional subspace is depicted that includes both the vectors **y** and **1**. *regression*: Using the same value to predict every response is equivalent to using a vector in span{**1**} to predict **y**. The sum of squared residuals equals the squared distance from **y** to its prediction vector, so the least-squares prediction vector is the orthogonal projection $\bar{y}\mathbf{1}$

## Visualizing location models

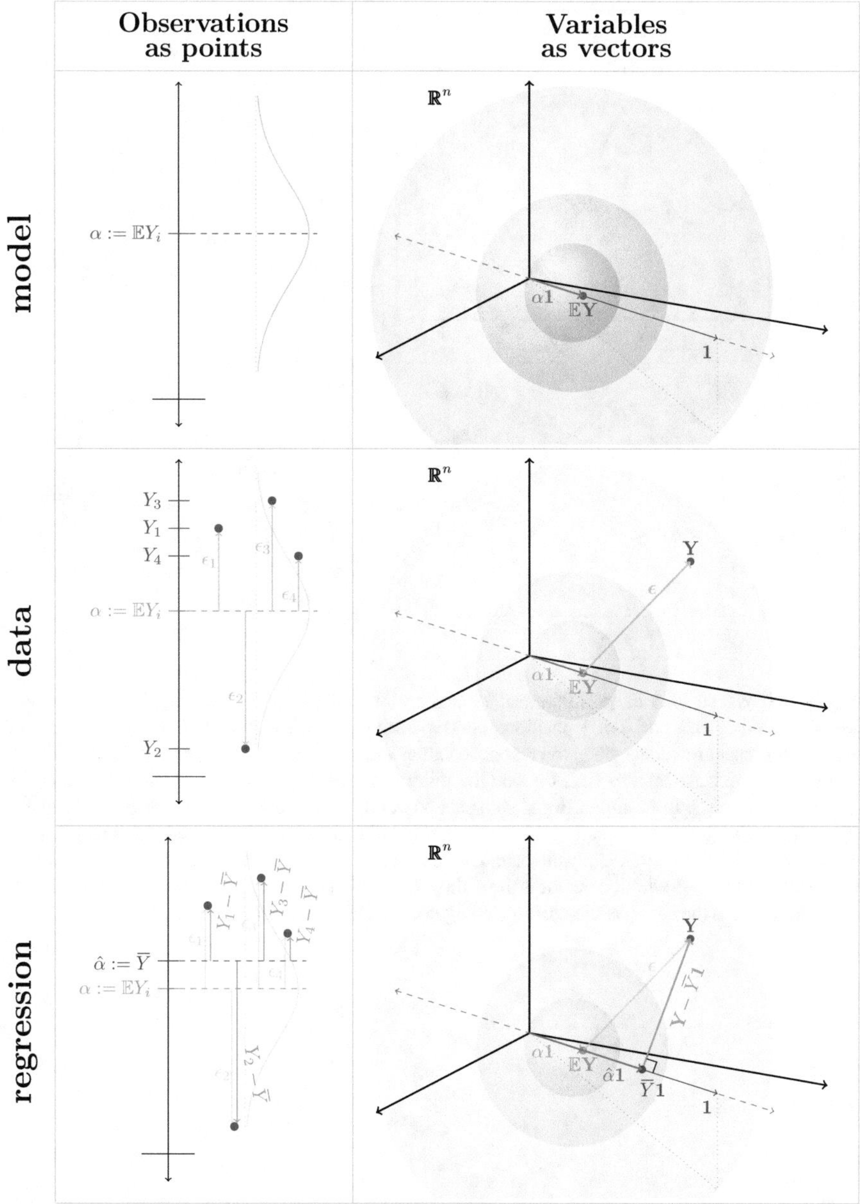


**Fig. 4.1** **Observations as points.** *model*: The response values share the same expectation $\alpha$. *data*: Arrows represent the random errors which "kick" the response values away from their expectation. *regression*: The least-squares point (average of the responses) can be considered an estimate of their expectation. **Variables as vectors.** *model*: A three-dimensional subspace is depicted that includes both the vectors **y** and **1**. If the responses share the same expectation, then the expectation vector is in span{**1**}. A density for the response is also depicted. *data*: The error random vector "kicks" the response vector away from its expectation. *regression*: The response is projected back into span{**1**} to arrive at the least-squares prediction vector $\hat{\alpha}\mathbf{1}$. The sample average $\hat{\alpha} = \overline{Y}$ can be considered an estimate of the common expectation $\alpha$

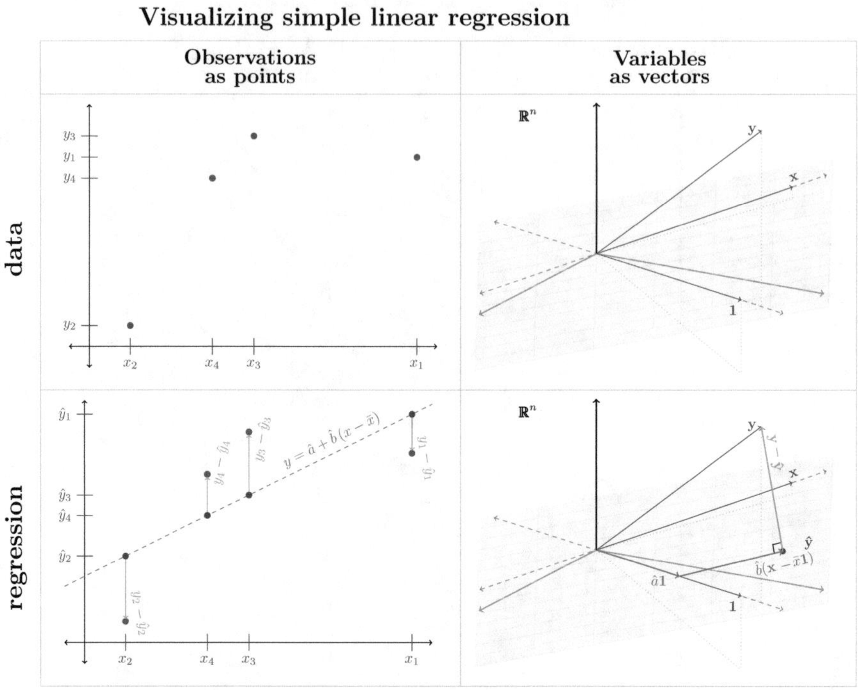


**Fig. 2.2 Observations as points.** *data*: A scatterplot draws points with heights equal to their response values and horizontal positions corresponding to their explanatory values. *regression*: Arrows represent residuals when the response values are predicted by the height of the dotted line. Depicted is the least-squares line, the one the minimizes the sum of squared residuals. **Variables as vectors.** *data*: A three-dimensional subspace is depicted that includes the vectors **y**, **1**, and **x**. *regression*: Using a line to predict the responses is equivalent to using a vector in span{**1**, **x**} to predict **y**. The sum of squared residuals equals the squared distance from **y** to its prediction vector, so the least-squares prediction vector is the orthogonal projection $\hat{\mathbf{y}}$. The coefficients of **1** and $\mathbf{x} - \bar{x}\mathbf{1}$ that lead to $\hat{\mathbf{y}}$ are the same as the corresponding coefficients of the least-squares line

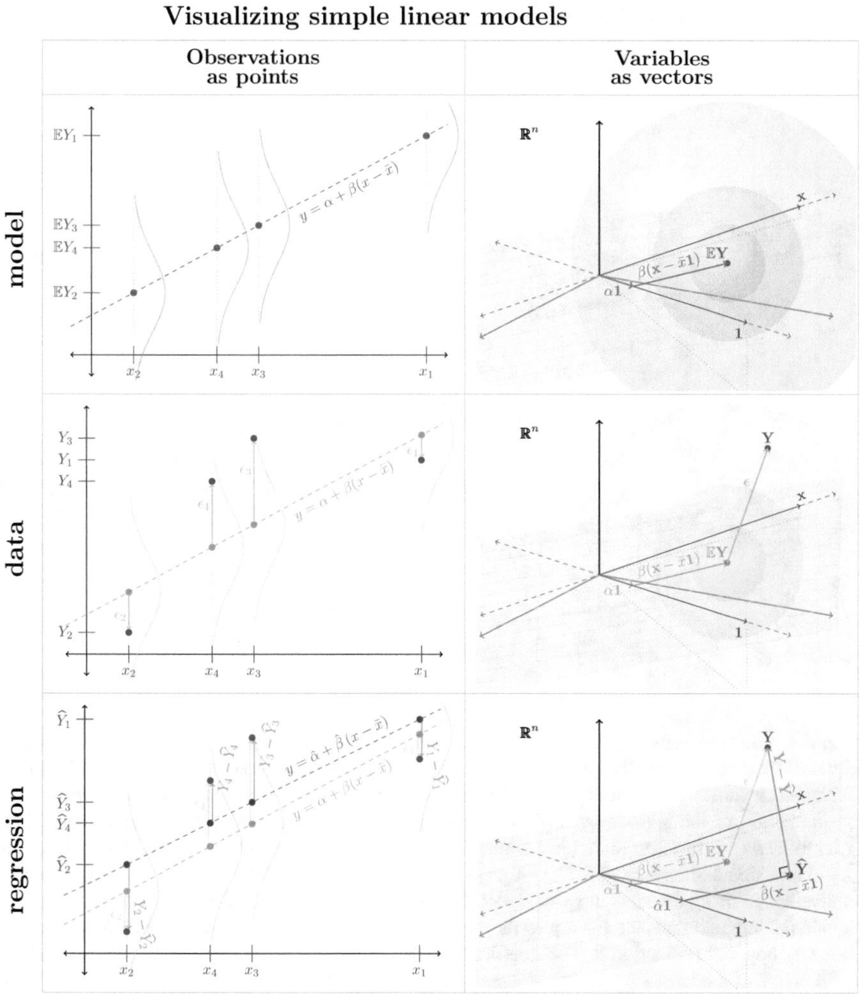


**Fig. 4.2 Observations as points.** *model*: The response values have expectations along the line $\alpha + \beta(x - \bar{x})$. *data*: Arrows represent the random errors which "kick" the response values away from their expectations. *regression*: The least-squares line provides estimates of their expectations. **Variables as vectors.** *model*: A three-dimensional subspace is depicted that includes the vectors **y**, **1**, and **x**. If the responses' expectations are along a line, then the expectation vector is in span$\{\mathbf{1}, \mathbf{x}\}$. A density for the response is also depicted. *data*: The error random vector "kicks" the response vector away from its expectation. *regression*: The response is projected back into span$\{\mathbf{1}, \mathbf{x}\}$ to arrive at the least-squares prediction vector $\hat{\alpha}\mathbf{1} + \hat{\beta}(\mathbf{x} - \bar{x}\mathbf{1})$. The coefficients $\hat{\alpha}$ and $\hat{\beta}$ can be considered estimates of the coefficients $\alpha$ and $\beta$ from the true expectations' line

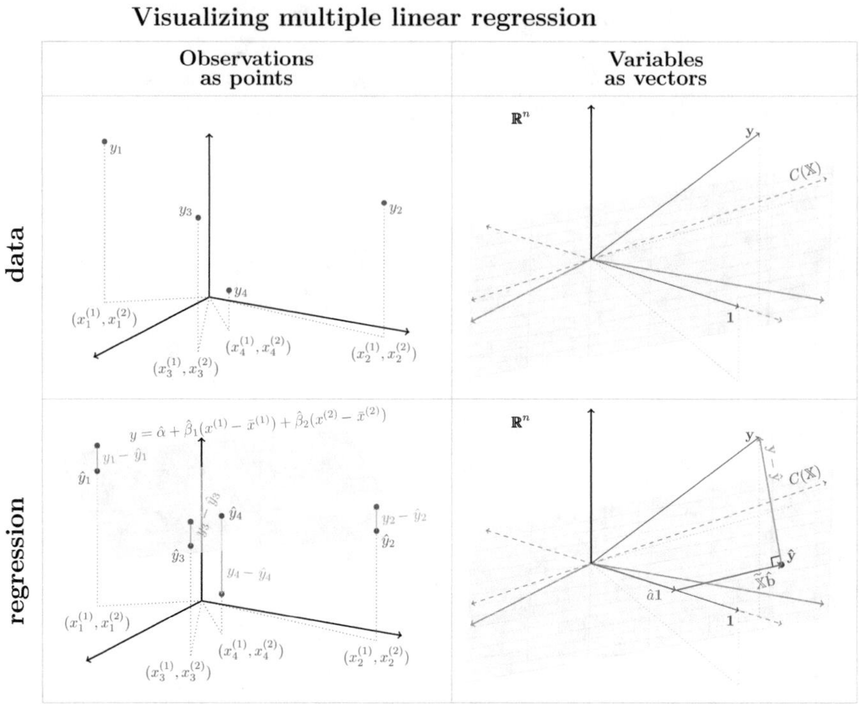


**Fig. 2.3 Observations as points.** *data*: A three-dimensional scatterplot draws points with heights equal to their response values and positions in the other two directions corresponding to their two explanatory values. *regression*: Arrows represent residuals when the response values are predicted by the height of the least-squares plane, the plane that minimizes the sum of squared residuals. **Variables as vectors.** *data*: A three-dimensional subspace is depicted that includes **y**, **1**, and a one-dimensional subspace of $C(\mathbb{X})$. *regression*: Using a hyperplane to predict the responses is equivalent to using a vector in span$\{\mathbf{1}, \mathbf{x}^{(1)}, \ldots, \mathbf{x}^{(m)}\}$ to predict **y**. The sum of squared residuals equals the squared distance from **y** to its prediction vector, so the least-squares prediction vector is the orthogonal projection $\hat{\mathbf{y}}$. The coefficients that lead to $\hat{\mathbf{y}}$ are the same as the corresponding coefficients of the least-squares hyperplane

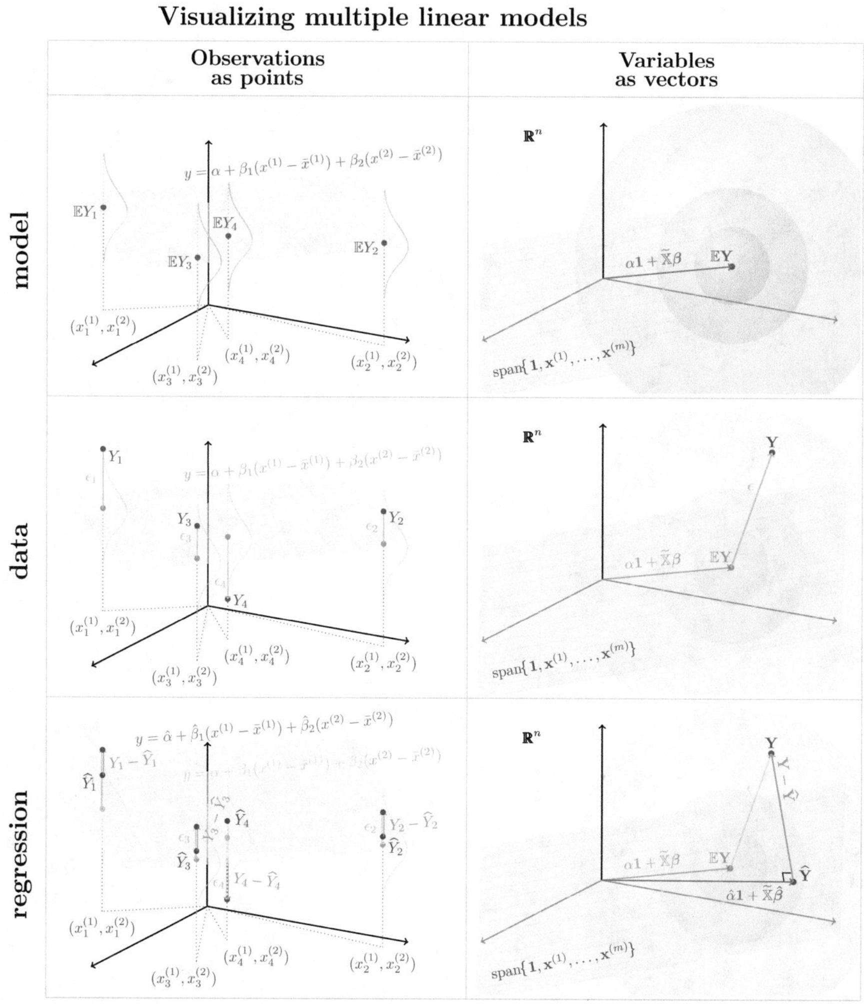


**Fig. 4.3 Observations as points.** *model*: The response values have expectations along a plane. *data*: Arrows represent the random errors which "kick" the response values away from their expectations. *regression*: The least-squares plane has the smallest possible sum of squared residuals. **Variables as vectors.** *model*: A three-dimensional subspace is depicted that includes **y** and a two-dimensional subspace of span$\{\mathbf{1}, \mathbf{x}^{(1)}, \ldots, \mathbf{x}^{(m)}\}$. If the responses' expectations are along a hyperplane, then the expectation vector is in span$\{\mathbf{1}, \mathbf{x}^{(1)}, \ldots, \mathbf{x}^{(m)}\}$. A density for the response is also depicted. *data*: The error random vector "kicks" the response vector away from its expectation. *regression*: The response is projected back into span$\{\mathbf{1}, \mathbf{x}^{(1)}, \ldots, \mathbf{x}^{(m)}\}$ to arrive at the least-squares prediction vector. Its coefficients can be considered estimates of the corresponding coefficients from the true expectations' hyperplane

# Appendix C
# Solutions

**Exercise 1.1**

*Solution* By definition, a vector space has the property that $\mathbf{v} - \mathbf{v} = \mathbf{0}$ for any vector $\mathbf{v}$. In particular, $\mathbf{0} - \mathbf{0} = \mathbf{0}$.

$$\begin{aligned} a\mathbf{0} &= a(\mathbf{0} - \mathbf{0}) \\ &= a\mathbf{0} - a\mathbf{0} \\ &= \mathbf{0}. \end{aligned}$$

We distributed scalar multiplication then used the $\mathbf{v} - \mathbf{v} = \mathbf{0}$ property again. ♦

**Exercise 1.2**

*Solution* The zero scalar satisfies $a - a = 0$ for every $a$ in the scalar field; in particular, $0 - 0 = 0$. We make this substitution, then distribute and invoke the fact that $\mathbf{v} - \mathbf{v}$ equals the zero vector $\mathbf{0}$ for any vector $\mathbf{v}$.

$$\begin{aligned} 0\mathbf{v} &= (0 - 0)\mathbf{v} \\ &= 0\mathbf{v} - 0\mathbf{v} \\ &= \mathbf{0}. \end{aligned}$$

♦

**Exercise 1.4**

*Solution* We will show that an arbitrary linear combination of $\mathbf{v}_1, \ldots, \mathbf{v}_m$ can also be represented as a linear combination of the altered vectors by adding and subtracting the appropriately scaled versions of $\mathbf{v}_m$.

$$\begin{aligned} b_1\mathbf{v}_1 + \ldots + b_{m-1}\mathbf{v}_{m-1} + b_m\mathbf{v}_m &= b_1(\mathbf{v}_1 + a_1\mathbf{v}_m) + \ldots + b_{m-1}(\mathbf{v}_{m-1} + a_{m-1}\mathbf{v}_m) \\ &\quad + (b_m - b_1a_1 - \ldots - b_{m-1}a_{m-1})\mathbf{v}_m. \end{aligned}$$


W. D. Brinda, *Visualizing Linear Models*,
https://doi.org/10.1007/978-3-030-64167-2

Similarly, an arbitrary linear combination of the altered vectors becomes a linear combination of $\mathbf{v}_1, \ldots, \mathbf{v}_m$ by distributing the coefficients and regrouping the terms.

$$b_1(\mathbf{v}_1 + a_1\mathbf{v}_1) + \ldots + b_{m-1}(\mathbf{v}_{m-1} + a_{m-1}\mathbf{v}_m) + b_m\mathbf{v}_m = b_1\mathbf{v}_1 + \ldots + b_{m-1}\mathbf{v}_{m-1}$$
$$+ (b_m + b_1a_1 + \ldots + b_{m-1}a_{m-1})\mathbf{v}_m.$$

♦

### Exercise 1.5

*Solution* First, consider that $\mathbb{T}$ and $\begin{bmatrix}\mathbf{v}_1 \cdots \mathbf{v}_m\end{bmatrix}$ would have to agree on where to map the standard basis vectors. We see that $\begin{bmatrix}\mathbf{v}_1 \cdots \mathbf{v}_m\end{bmatrix}$ maps $\mathbf{e}_1 := (1, 0, \ldots, 0)$ to $\mathbf{v}_1$, so $\mathbf{v}_1$ must be $\mathbb{T}\mathbf{e}_1$. Likewise $\mathbf{v}_2$ would need to $\mathbb{T}\mathbf{e}_2$, and so on. Let us check that this proposal $\begin{bmatrix}\mathbb{T}\mathbf{e}_1 \ldots \mathbb{T}\mathbf{e}_m\end{bmatrix}\mathbf{b}$ is the same as $\mathbb{T}\mathbf{b}$ for an arbitrary vector $\mathbf{b}$.

$$\begin{aligned}
\begin{bmatrix}\mathbb{T}\mathbf{e}_1 \ldots \mathbb{T}\mathbf{e}_m\end{bmatrix}\mathbf{b} &= \begin{bmatrix}\mathbb{T}\mathbf{e}_1 \ldots \mathbb{T}\mathbf{e}_m\end{bmatrix}(b_1\mathbf{e}_1 + \ldots + b_m\mathbf{e}_m) \\
&= b_1\begin{bmatrix}\mathbb{T}\mathbf{e}_1 \ldots \mathbb{T}\mathbf{e}_m\end{bmatrix}\mathbf{e}_1 + \ldots + b_m\begin{bmatrix}\mathbb{T}\mathbf{e}_1 \ldots \mathbb{T}\mathbf{e}_m\end{bmatrix}\mathbf{e}_m \\
&= b_1\mathbb{T}\mathbf{e}_1 + \ldots + b_m\mathbb{T}\mathbf{e}_m \\
&= \mathbb{T}(b_1\mathbf{e}_1 + \ldots + b_m\mathbf{e}_m) \\
&= \mathbb{T}\mathbf{b}.
\end{aligned}$$

♦

### Exercise 1.7

*Solution* Assume there exist scalars $b_1, \ldots, b_m$ for which $b_1\mathbf{v}_1 + \ldots + b_m\mathbf{v}_m = 0$ with $b_1 \neq 0$ (without loss of generality). Then the rearranged equation

$$\mathbf{v}_1 = \left(-\tfrac{b_2}{b_1}\right)\mathbf{v}_2 + \ldots + \left(-\tfrac{b_m}{b_1}\right)\mathbf{v}_m$$

shows that $\mathbf{v}_1$ is a linear combination of the other vectors, contradicting the assumption of linear independence. ♦

### Exercise 1.8

*Solution* Assume that $\mathbf{v}_1, \ldots, \mathbf{v}_m$ are not linearly independent; in particular, and without loss of generality, assume $\mathbf{v}_m = c_1\mathbf{v}_1 + \ldots + c_{m-1}\mathbf{v}_{m-1}$ for some scalars $c_1, \ldots, c_{m-1}$. Then subtracting $\mathbf{v}_m$ from both sides, we see that

$$\mathbf{0} = c_1\mathbf{v}_1 + \ldots + c_{m-1}\mathbf{v}_{m-1} + (-1)\mathbf{v}_m$$

provides a linear combination of zero in which not all of the scalar coefficients are zeros. ♦

**Exercise 1.10**

*Solution* Trivially $\mathbf{c} = \mathbf{b} + (\mathbf{c} - \mathbf{b})$; we show that the second term is in the null space.

$$\begin{aligned}\begin{bmatrix}\mathbf{v}_1 & \cdots & \mathbf{v}_m\end{bmatrix}(\mathbf{c} - \mathbf{b}) &= \begin{bmatrix}\mathbf{v}_1 & \cdots & \mathbf{v}_m\end{bmatrix}\mathbf{c} - \begin{bmatrix}\mathbf{v}_1 & \cdots & \mathbf{v}_m\end{bmatrix}\mathbf{b} \\ &= \mathbf{0}\end{aligned}$$

because the two linear combinations are equal. ♦

**Exercise 1.11**

*Solution* With $\mathcal{N}$ representing the null space of $\begin{bmatrix}\mathbf{v}_1 & \cdots & \mathbf{v}_m\end{bmatrix}$, the set of coefficient vectors producing $\begin{bmatrix}\mathbf{v}_1 & \cdots & \mathbf{v}_m\end{bmatrix}\mathbf{b}$ is exactly $\{\mathbf{b} + \mathbf{z} : \mathbf{z} \in \mathcal{N}\}$; this set has a single element ($\mathbf{b}$) if and only if the null space has only the zero vector. We have already established that this condition is equivalent to linear independence. ♦

**Exercise 1.12**

*Solution* By linear independence, no linear combination of $\mathbf{v}_2, \ldots, \mathbf{v}_m$ equals $\mathbf{v}_1 \in \mathcal{V}$. Therefore, $\mathbf{v}_2, \ldots, \mathbf{v}_m$ do not span $\mathcal{V}$ and thus do not satisfy the definition of a basis. ♦

**Exercise 1.13**

*Solution* Any $\mathbf{w}$ in $\mathcal{V}$ has a unique representation as $b_1\mathbf{v}_1 + \ldots b_m\mathbf{v}_m$ for some scalar coefficients. Assume every one of the basis vectors were in $\mathcal{S}$. Because $\mathcal{S}$ is a subspace it contains all linear combinations of its vectors which would include $\mathbf{w}$. Therefore, if $\mathcal{S}$ is a subspace of $\mathcal{V}$ that includes an entire basis for $\mathcal{V}$, then it would have to be equal to $\mathcal{V}$. ♦

**Exercise 1.14**

*Solution* Suppose that $\{\mathbf{w}_1, \ldots, \mathbf{w}_k\}$ is also a basis for $\mathcal{V}$ and that $k < m$. The set $\{\mathbf{v}_2, \ldots, \mathbf{v}_m\}$ does not span $\mathcal{V}$, so there must be some vector in $\{\mathbf{w}_1, \ldots, \mathbf{w}_k\}$ that is not in the span of $\{\mathbf{v}_2, \ldots, \mathbf{v}_m\}$. Assume without loss of generality that $\mathbf{w}_1$ is not in the span of $\{\mathbf{v}_2, \ldots, \mathbf{v}_m\}$. Then $\{\mathbf{w}_1, \mathbf{v}_2, \ldots, \mathbf{v}_m\}$ is linearly independent. By repeating this logic with $\mathbf{v}_2$ and continuing through $\mathbf{v}_k$, we end up claiming that $\{\mathbf{w}_1, \ldots, \mathbf{w}_k, \mathbf{v}_{k+1}, \ldots, \mathbf{v}_m\}$ is a linearly independent set. However, since $\{\mathbf{w}_1, \ldots, \mathbf{w}_k\}$ is a basis for $\mathcal{V}$, the remaining vectors $\{\mathbf{v}_{k+1}, \ldots, \mathbf{v}_m\} \subset \mathcal{V}$ have to be in their span which is a contradiction. This proves that one basis for $\mathcal{V}$ cannot be larger than another. ♦

**Exercise 1.15**

*Solution* Let $V := \{\mathbf{v}_1, \ldots, \mathbf{v}_m\}$ be a set of linearly independent vectors in $\mathcal{V}$, and let $W := \{\mathbf{w}_1, \ldots, \mathbf{w}_m\}$ be a basis for $\mathcal{V}$. Assume that the span of $V$ is not $\mathcal{V}$. From Exercise 1.13 there exists at least one vector in $W$ that is not in the span of $V$; without loss of generality, let it be $\mathbf{w}_1$. By continuing to add one vector at a time from $V$ to $W$ in this way, you would maintain linear independence and you would eventually produce a set that does have $\mathcal{V}$ as its span. But that set will have more than $m$ vectors, which contradicts the fact from Exercise 1.14 that every basis for $\mathcal{V}$ has $m$ vectors. ♦

**Exercise 1.16**

*Solution* We will describe a constructive proof. Take any vector of $\mathcal{S}$ and call it $\mathbf{v}_1$. If $\mathcal{S}$ equals the span of $\mathbf{v}_1$, then $\mathbf{v}_1$ is a basis. Otherwise, take a vector from $\mathcal{S}$ that is outside of the span of $\mathbf{v}_1$ and call it $\mathbf{v}_2$. If $\mathcal{S}$ equals the span of $\{\mathbf{v}_1, \mathbf{v}_2\}$, then they form a basis. Otherwise, continue to repeat this process of adding one more linearly independent vector at a time until the vectors span $\mathcal{S}$. The algorithm is guaranteed to terminate with no more than $n$ vectors, because any set of $n$ linearly independent vectors is a basis for $\mathcal{V}$ according to Exercise 1.15. ♦

**Exercise 1.22**

*Solution* We know that $\mathbf{w} = \mathbb{T}\mathbf{v}$ for some $\mathbf{v}$; by making this substitution,

$$\begin{aligned} \mathbb{T}\mathbb{T}^{-1}\mathbf{w} &= \mathbb{T}\mathbb{T}^{-1}\mathbb{T}\mathbf{v} \\ &= \mathbb{T}\mathbf{v} \\ &= \mathbf{w}. \end{aligned}$$

♦

**Exercise 1.23**

*Solution* As an inverse function, $\mathbb{T}^{-1}$ maps from $\mathbb{T}$'s range to its domain. $\mathbf{w}_1$ and $\mathbf{w}_2$ are in the range of $\mathbb{T}$, so we know that they can be represented by $\mathbf{w}_1 = \mathbb{T}\mathbf{v}_1$ and $\mathbf{w}_2 = \mathbb{T}\mathbf{v}_2$ for some vectors $\mathbf{v}_1$ and $\mathbf{v}_2$.

$$\begin{aligned} \mathbb{T}^{-1}(a_1\mathbf{w}_1 + a_2\mathbf{w}_2) &= \mathbb{T}^{-1}(a_1\mathbb{T}\mathbf{v}_1 + a_2\mathbb{T}\mathbf{v}_2) \\ &= \mathbb{T}^{-1}\mathbb{T}(a_1\mathbf{v}_1 + a_2\mathbf{v}_2) \\ &= a_1\mathbf{v}_1 + a_2\mathbf{v}_2 \\ &= a_1\mathbb{T}^{-1}\mathbf{w}_1 + a_2\mathbb{T}^{-1}\mathbf{w}_2. \end{aligned}$$

♦

**Exercise 1.25**

*Solution* The complex conjugate of a sum is equal to the sum of the complex conjugates, so

$$\begin{aligned} \langle \mathbf{v}, \mathbf{w} + \mathbf{x} \rangle &= \overline{\langle \mathbf{w} + \mathbf{x}, \mathbf{v} \rangle} \\ &= \overline{\langle \mathbf{w}, \mathbf{v} \rangle} + \overline{\langle \mathbf{x}, \mathbf{v} \rangle} \\ &= \langle \mathbf{v}, \mathbf{w} \rangle + \langle \mathbf{v}, \mathbf{x} \rangle. \end{aligned}$$

♦

### Exercise 1.26

*Solution* The complex conjugate of a product is equal to the product of the complex conjugates, and the complex conjugate of a real number is itself, so

$$\begin{aligned}\langle \mathbf{v}, a\mathbf{w}\rangle &= \overline{\langle a\mathbf{w}, \mathbf{v}\rangle}\\ &= \overline{a\langle \mathbf{w}, \mathbf{v}\rangle}\\ &= \overline{a}\overline{\langle \mathbf{w}, \mathbf{v}\rangle}\\ &= a\langle \mathbf{v}, \mathbf{w}, \rangle.\end{aligned}$$

♦

### Exercise 1.28

*Solution* From steps in our solution to Exercise 1.26, we can realize that $\langle \mathbf{x}, a\mathbf{y}\rangle = \overline{a}\langle \mathbf{x}, \mathbf{y}\rangle$. Using the definition of norm,

$$\begin{aligned}\|a\mathbf{v}\| &= \sqrt{\langle a\mathbf{v}, a\mathbf{v}\rangle}\\ &= \sqrt{a\overline{a}}\sqrt{\langle \mathbf{v}, \mathbf{v}\rangle}\\ &= |a|\|\mathbf{v}\|.\end{aligned}$$

Note that with $a = b + ic$, the squared absolute value $|a|^2$ is defined to be $b^2 + c^2$. ♦

### Exercise 1.29

*Solution* Without loss of generality, we will consider whether or not $\mathbf{v}_m$ can be equal to some linear combination $b_1\mathbf{v}_1 + \ldots + b_{m-1}\mathbf{v}_{m-1}$. The inner product of this linear combination with $\mathbf{v}_m$ equals

$$\begin{aligned}\langle b_1\mathbf{v}_1 + \ldots + b_{m-1}\mathbf{v}_{m-1}, \mathbf{v}_m\rangle &= b_1 \underbrace{\langle \mathbf{v}_1, \mathbf{v}_m\rangle}_{0} + \ldots + b_{m-1} \underbrace{\langle \mathbf{v}_{m-1}, \mathbf{v}_m\rangle}_{0}\\ &= 0.\end{aligned}$$

But the inner product of $\mathbf{v}_m$ with itself is equal to its squared length, which must be greater than zero by the assumption that $\mathbf{v}_m$ is not the zero vector. Therefore, no such linear combination can be equal to $\mathbf{v}_m$. ♦

**Exercise 1.30**

*Solution* Subtracting, $\langle \mathbf{y}, \mathbf{v} \rangle$ from both sides of the assumption,

$$\begin{aligned} 0 &= \langle \mathbf{x}, \mathbf{v} \rangle - \langle \mathbf{y}, \mathbf{v} \rangle \\ &= \langle \mathbf{x} - \mathbf{y}, \mathbf{v} \rangle \end{aligned}$$

for all $\mathbf{v}$. In particular, substitute $\mathbf{x} - \mathbf{y}$ for $\mathbf{v}$ to see that $\|\mathbf{x} - \mathbf{y}\|^2 = 0$ which implies that $\mathbf{x} - \mathbf{y} = \mathbf{0}$, i.e. $\mathbf{x} = \mathbf{y}$. ♦

**Exercise 1.45**

*Solution* Let $B$ be a basis for $\mathcal{S}$; it contains $m$ vectors. Suppose there exist more than $d - m$ linearly independent vectors in $\mathcal{S}^\perp$. All of them are also linearly independent of $B$, so the two bases taken together would contain a total of *more than d* linearly independent vectors in $\mathcal{V}$ which is impossible. On the other hand, suppose $\mathcal{S}^\perp$ has a basis of fewer than $d - m$ vectors. Then that basis, taken together with $B$ would have fewer than $d$ vectors, so it would not span $\mathcal{V}$. However, this cannot be true because we have seen than *any* vector in $\mathcal{V}$ can be represented as the sum of a vector in $\mathcal{S}$ and a vector in $\mathcal{S}^\perp$. ♦

**Exercise 1.49**

*Solution* The operator $\mathbb{H}_1 - \mathbb{H}_0$ evaluated at $\mathbf{y}$ has the value $\mathbb{H}_1\mathbf{y} - \mathbb{H}_0\mathbf{y}$. From our discussion regarding Eq. 1.4, we know that this is precisely the orthogonal projection of $\mathbf{y}$ onto the orthogonal complement of $\mathcal{S}_0$ within $\mathcal{S}_1$. ♦

**Exercise 1.50**

*Solution* In our discussion regarding Eq. 1.4, we realized that the orthogonal projection of $\mathbf{y}$ onto $\mathcal{S}_0$ is the same as the orthogonal projection of $\mathbb{H}_1\mathbf{y}$ onto $\mathcal{S}_0$. In other words, it does not matter which order you compose the operators, you end up at $\mathbb{H}_0\mathbf{y}$ either way. ♦

**Exercise 1.51**

*Solution* The $i$th diagonal entry of $\mathbb{ML}$ is $\sum_{j=1}^{m} \mathbb{M}_{i,j}\mathbb{L}_{j,i}$. We express the trace as the sum of these diagonals then reverse the order of the summations.

$$\begin{aligned} \operatorname{tr} \mathbb{ML} &= \sum_{i=1}^{n}\sum_{j=1}^{m} \mathbb{M}_{i,j}\mathbb{L}_{j,i} \\ &= \sum_{j=1}^{m}\sum_{i=1}^{n} \mathbb{L}_{j,i}\mathbb{M}_{i,j} \\ &= \operatorname{tr} \mathbb{LM}. \end{aligned}$$

♦

### Exercise 1.52

*Solution* The $(i, j)$ entry of $(\mathbb{M}\mathbb{L})^T$ is the $(j, i)$ entry of $\mathbb{M}\mathbb{L}$, which is $\sum_k \mathbb{M}_{j,k}\mathbb{L}_{k,i}$. The $(i, j)$ entry of $\mathbb{L}^T\mathbb{M}^T$ works out to be the same:

$$\sum_k (\mathbb{L}^T)_{i,k}(\mathbb{M}^T)_{k,j} = \sum_k \mathbb{L}_{k,i}\mathbb{M}_{j,k}.$$

♦

### Exercise 1.53

*Solution* Letting $\mathbf{u}_1, \ldots, \mathbf{u}_m$ denote the columns,

$$\begin{aligned}\mathbb{U}^T\mathbb{U} &= \begin{bmatrix} - & \mathbf{u}_1 & - \\ & \vdots & \\ - & \mathbf{u}_m & - \end{bmatrix}\begin{bmatrix} | & & | \\ \mathbf{u}_1 & \ldots & \mathbf{u}_m \\ | & & | \end{bmatrix} \\ &= \begin{bmatrix} \mathbf{u}_1^T\mathbf{u}_1 & \cdots & \mathbf{u}_1^T\mathbf{u}_n \\ \vdots & \ddots & \vdots \\ \mathbf{u}_n^T\mathbf{u}_1 & \cdots & \mathbf{u}_n^T\mathbf{u}_n \end{bmatrix} \\ &= \begin{bmatrix} 1 & & \\ & \ddots & \\ & & 1 \end{bmatrix}\end{aligned}$$

with every off-diagonal entry equal to zero. ♦

### Exercise 1.55

*Solution* Let $\mathbb{M}$ be a square matrix. First, assume it is invertible. Then what linear combinations satisfy $\mathbb{M}\mathbf{b} = \mathbf{0}$? Multiplying both sides by the inverse, we see that the coefficients $\mathbf{b} = \mathbb{M}^{-1}\mathbf{0}$ must be the zero vector.

Next, suppose the $n$ columns of $\mathbb{M}$ are linearly independent. Then for each canonical basis vector $\mathbf{e}_j$, there is some coefficient vector $\mathbf{b}_j$ such that $\mathbb{M}\mathbf{b}_j = \mathbf{e}_j$. The matrix with these coefficient vectors as its columns is the inverse of $\mathbb{M}$.

$$\begin{aligned}\mathbb{M}\begin{bmatrix} | & & | \\ \mathbf{b}_1 & \cdots & \mathbf{b}_n \\ | & & | \end{bmatrix} &= \begin{bmatrix} | & & | \\ \mathbb{M}\mathbf{b}_1 & \cdots & \mathbb{M}\mathbf{b}_n \\ | & & | \end{bmatrix} \\ &= \underbrace{\begin{bmatrix} | & & | \\ \mathbf{e}_1 & \cdots & \mathbf{e}_n \\ | & & | \end{bmatrix}}_{\mathbb{I}_n}.\end{aligned}$$

♦

## Exercise 1.56

*Solution* Let us check what would be required for an arbitrary vector $\mathbf{w}$ to be an eigenvector for $\mathbb{M}$. We can express $\mathbf{w}$ with respect to the eigenvector basis:

$$\begin{aligned}\mathbb{M}\mathbf{w} &= \mathbb{M}(\langle \mathbf{w}, \mathbf{q}_1\rangle \mathbf{q}_1 + \ldots + \langle \mathbf{w}, \mathbf{q}_n\rangle \mathbf{q}_n)\\ &= \langle \mathbf{w}, \mathbf{q}_1\rangle \mathbb{M}\mathbf{q}_1 + \ldots + \langle \mathbf{w}, \mathbf{q}_n\rangle \mathbb{M}\mathbf{q}_n\\ &= \langle \mathbf{w}, \mathbf{q}_1\rangle \lambda_1\mathbf{q}_1 + \ldots + \langle \mathbf{w}, \mathbf{q}_n\rangle \lambda_n\mathbf{q}_n.\end{aligned}$$

This is only proportional to $\mathbf{w} = \langle \mathbf{w}, \mathbf{q}_1\rangle \mathbf{q}_1 + \ldots + \langle \mathbf{w}, \mathbf{q}_n\rangle \mathbf{q}_n$ if all of the non-zero terms share the same eigenvalue coefficient. That coefficient, which is one of $\lambda_1, \ldots, \lambda_n$, would be the eigenvalue of $\mathbf{w}$. ♦

## Exercise 1.63

*Solution* The columns of $\mathbb{M}$ are linearly independent if and only if $C(\mathbb{M})$ is $m$-dimensional; that is what we will check.

First, suppose $m \leq n$, and let $\mathbb{V}$ have the singular value decomposition

$$\mathbb{V} = \sigma_1\mathbf{u}_1\mathbf{v}_1^T + \ldots + \sigma_m\mathbf{u}_m\mathbf{v}_m^T.$$

Then $\mathbb{V}^T\mathbb{V}$ has the spectral decomposition

$$\mathbb{V}^T\mathbb{V} = \sigma_1^2\mathbf{v}_1\mathbf{v}_1^T + \ldots + \sigma_m^2\mathbf{v}_m\mathbf{v}_m^T.$$

We know that $\mathbb{V}^T\mathbb{V}$ is invertible if and only if its eigenvalues $\sigma_1^2, \ldots, \sigma_m^2$ are positive. The column space of $\mathbb{V}$ is a linear combination of all of $\mathbf{u}_1, \ldots, \mathbf{u}_m$ if and only if none of the singular values $\sigma_1, \ldots, \sigma_m$ is zero. These conditions are the same.

Otherwise, if $n < m$, then the singular decomposition cannot possibly represent a linear combination of $m$ column vectors, so $\mathbb{M}$ cannot have linearly independent. Similarly, the squared singular values cannot account for the $m$ positive eigenvalues that $\mathbb{V}^T\mathbb{V}$ would need to be invertible. ♦

## Exercise 1.64

*Solution* First, assume that $\mathbb{M}$ is symmetric. For an arbitrary $\mathbf{v}, \mathbf{w} \in \mathbb{R}^n$, we can use a spectral decomposition of $\mathbb{M}$ to see that

$$\begin{aligned}\langle \mathbf{v}, \mathbb{M}\mathbf{w}\rangle &= \mathbf{v}^T\mathbb{M}\mathbf{w}\\ &= \mathbf{v}^T(\lambda_1\mathbf{q}_1\mathbf{q}_1^T + \ldots + \lambda_n\mathbf{q}_n\mathbf{q}_n^T)\mathbf{w}\\ &= \lambda_1(\mathbf{v}^T\mathbf{q}_1)(\mathbf{q}_1^T\mathbf{w}) + \ldots + \lambda_n(\mathbf{v}^T\mathbf{q}_n)(\mathbf{q}_n^T\mathbf{w})\\ &= \lambda_1(\mathbf{w}^T\mathbf{q}_1)(\mathbf{q}_1^T\mathbf{v}) + \ldots + \lambda_n(\mathbf{w}^T\mathbf{q}_n)(\mathbf{q}_n^T\mathbf{v})\end{aligned}$$

$$
\begin{aligned}
&= \mathbf{w}^T(\lambda_1 \mathbf{q}_1 \mathbf{q}_1^T + \ldots + \lambda_n \mathbf{q}_n \mathbf{q}_n^T)\mathbf{v} \\
&= \mathbf{w}^T \mathbb{M}\mathbf{v} \\
&= \langle \mathbf{w}, \mathbb{M}\mathbf{v} \rangle.
\end{aligned}
$$

Next, suppose $\langle \mathbf{v}, \mathbb{M}\mathbf{w} \rangle = \langle \mathbb{M}\mathbf{v}, \mathbf{w} \rangle$ for every $\mathbf{v}, \mathbf{w} \in \mathbb{R}^n$. In particular, apply two canonical basis vectors $\mathbf{e}_i$ and $\mathbf{e}_j$. The vector $\mathbb{M}\mathbf{e}_j$ is the $j$th column of $\mathbb{M}$, so $\langle \mathbf{e}_i, \mathbb{M}\mathbf{e}_j \rangle$ is the $(i, j)$-entry of $\mathbb{M}$. By our assumption, it is equal to $\langle \mathbf{e}_j, \mathbb{M}\mathbf{e}_i \rangle$ which is the $(j, i)$-entry of $\mathbb{M}$ which shows that $\mathbb{M}$ must be symmetric. ♦

**Exercise 1.65**

*Solution*

$$
\begin{aligned}
\langle \mathbb{H}\mathbf{v}, \mathbf{w} \rangle &= \langle \mathbb{H}\mathbf{v}, \mathbb{H}\mathbf{w} + (\mathbf{w} - \mathbb{H}\mathbf{w}) \rangle \\
&= \langle \mathbb{H}\mathbf{v}, \mathbb{H}\mathbf{w} \rangle + \langle \mathbb{H}\mathbf{v}, \mathbf{w} - \mathbb{H}\mathbf{w} \rangle.
\end{aligned}
$$

Because $\mathbb{H}\mathbf{w}$ is in the space that $\mathbb{H}$ projects onto while $\mathbf{w} - \mathbb{H}\mathbf{w}$ is orthogonal to it, the second term is zero. Similarly,

$$
\begin{aligned}
\langle \mathbf{v}, \mathbb{H}\mathbf{w} \rangle &= \langle \mathbb{H}\mathbf{v} + (\mathbf{v} - \mathbb{H}\mathbf{v}), \mathbb{H}\mathbf{w} \rangle \\
&= \langle \mathbb{H}\mathbf{v}, \mathbb{H}\mathbf{w} \rangle + \underbrace{\langle \mathbf{v} - \mathbb{H}\mathbf{v}, \mathbb{H}\mathbf{w} \rangle}_{0}.
\end{aligned}
$$

Both $\langle \mathbf{v}, \mathbb{H}\mathbf{w} \rangle$ and $\langle \mathbb{H}\mathbf{v}, \mathbf{w} \rangle$ simplify to $\langle \mathbb{H}\mathbf{v}, \mathbb{H}\mathbf{w} \rangle$, so they are equal to each other.

It is also clear from the formula derived in Exercise 1.71 that orthogonal projection matrices are symmetric. However, I prefer the argument used here because it is more readily extended to orthogonal projection *operators*. ♦

**Exercise 1.72**

*Solution* Let $\mathbb{M} = \mathbb{U}\mathbb{S}\mathbb{V}^T$ be a singular value decomposition for which $\mathbb{S}$ is square and has only positive values along its diagonal.

$$
\begin{aligned}
\mathbb{M}^{-}\mathbb{M} &= \mathbb{V}\mathbb{S}^{-1}\mathbb{U}^T\mathbb{U}\mathbb{S}\mathbb{V}^T \\
&= \mathbb{V}\mathbb{V}^T.
\end{aligned}
$$

Exercise 1.71 indicates that the orthogonal projection matrix onto $C(\mathbb{V})$ is $\mathbb{V}(\mathbb{V}^T\mathbb{V})^{-1}\mathbb{V}^T$ which simplifies to $\mathbb{V}\mathbb{V}^T$ because $\mathbb{V}^T\mathbb{V}$ is the identity.

It only remains to establish that $C(\mathbb{V})$ is the row space of $\mathbb{M}$. Letting the entries of $\mathbf{w}$ be the coefficients of the linear combination,

$$
\begin{aligned}
\mathbf{w}^T\mathbb{M} &= \mathbf{w}^T(\sigma_1 \mathbf{u}_1 \mathbf{v}_1^T + \ldots \sigma_d \mathbf{u}_d \mathbf{v}_d^T) \\
&= \sigma_1 \langle \mathbf{w}, \mathbf{u}_1 \rangle \mathbf{v}_1^T + \ldots + \sigma_d \langle \mathbf{w}, \mathbf{u}_d \rangle \mathbf{v}_d^T.
\end{aligned}
$$

Any linear combination of $\mathbf{v}_1^T, \ldots, \mathbf{v}_d^T$ can be produced by the appropriate choice of $\mathbf{w}$, but no vector outside of their span can be produced. ♦

### Exercise 1.73

*Solution* If $\mathbb{M} \in \mathbb{R}^{n \times n}$ is invertible, then its row space is $\mathbb{R}^n$. Exercise 1.72 implies that $\mathbb{M}^- \mathbb{M}\mathbf{w} = \mathbf{w}$ for every $\mathbf{w} \in \mathbb{R}^n$. $\mathbb{M}^-$ must be the inverse according to Exercise 1.22. ♦

### Exercise 1.74

*Solution* From Sect. 1.5, we know that every solution can be represented as $\hat{\mathbf{b}} + \mathbf{w}$ for some $\mathbf{w}$ in the null space of $\mathbb{M}^T\mathbb{M}$. Let $\mathbb{M}^T\mathbb{M}$ have spectral decomposition

$$\mathbb{M}^T\mathbb{M} = \sigma_1^2 \mathbf{v}_1 \mathbf{v}_1^T + \ldots + \sigma_d^2 \mathbf{v}_d \mathbf{v}_d^T + (0)\mathbf{v}_{d+1}\mathbf{v}_{d+1}^T + \ldots + (0)\mathbf{v}_m \mathbf{v}_m^T.$$

with positive $\sigma_1^2, \ldots, \sigma_d^2$. It is clear that the null space is exactly the span of $\{\mathbf{v}_{d+1}, \ldots, \mathbf{v}_m\}$. On the other hand, by definition of generalized inverse, $\hat{\mathbf{b}}$ is in the span of $\{\mathbf{v}_1, \ldots, \mathbf{v}_d\}$. The squared norm of any solution $\hat{\mathbf{b}} + \mathbf{w}$ is $\|\hat{\mathbf{b}}\|^2 + \|\mathbf{w}\|^2$, so the solution of smallest norm is $\hat{\mathbf{b}}$. ♦

### Exercise 1.75

*Solution* Let $\mathbb{M}$ have singular value decomposition $\mathbb{U}\mathbb{S}\mathbb{V}^T$ where $\mathbb{S}$ is a square matrix with strictly positive diagonals. Then

$$\begin{aligned}
\mathbb{M}\mathbb{M}^-\mathbb{M} &= \mathbb{U}\mathbb{S}\underbrace{\mathbb{V}^T\mathbb{V}}_{\mathbb{I}}\mathbb{S}^{-1}\underbrace{\mathbb{U}^T\mathbb{U}}_{\mathbb{I}}\mathbb{S}\mathbb{V}^T \\
&= \mathbb{U}\mathbb{S}\mathbb{V}^T \\
&= \mathbb{M}.
\end{aligned}$$

♦

### Exercise 2.5

*Solution*

$$\begin{aligned}
\rho_{\mathbf{x},\mathbf{y}} &:= \frac{\sigma_{\mathbf{x},\mathbf{y}}}{\sigma_{\mathbf{x}}\sigma_{\mathbf{y}}} \\
&= \frac{(1/n)\langle \mathbf{x} - \bar{x}\mathbf{1}, \mathbf{y} - \bar{y}\mathbf{1}\rangle}{(\sqrt{1/n}\|\mathbf{x} - \bar{x}\mathbf{1}\|)(\sqrt{1/n}\|\mathbf{y} - \bar{y}\mathbf{1}\|)} \\
&= \frac{\langle \mathbf{x} - \bar{x}\mathbf{1}, \mathbf{y} - \bar{y}\mathbf{1}\rangle}{\|\mathbf{x} - \bar{x}\mathbf{1}\|\,\|\mathbf{y} - \bar{y}\mathbf{1}\|}.
\end{aligned}$$

♦

**Exercise 3.1**

*Solution* The union of $E_1$ and $E_2$ is the same as the union of the disjoint sets $E_1$ and $E_2/E_1$ (the part of $E_2$ that is not in $E_1$). With $\mathbb{P}$ mapping each event to its probability,

$$\begin{aligned}\mathbb{P}(E_1 \cup E_2) &= \mathbb{P}[E_1 \cup (E_2/E_1)]\\ &= \mathbb{P}E_1 + \mathbb{P}(E_2/E_1)\\ &\leq \mathbb{P}E_1 + \mathbb{P}E_2.\end{aligned}$$

To understand the last step, realize that $E_2$ can be represented as the disjoint union $E_2 = (E_2/E_1) \cup (E_2 \cap E_1)$. ♦

**Exercise 3.2**

*Solution* Let $\mathbf{X} = Y + iZ$ with random variables $Y$ and $Z$, and let $a = b + ic$ be a complex number.

$$\begin{aligned}\mathbb{E}\langle \mathbf{X}, a\rangle &= \mathbb{E}\langle Y + iZ, b + ic\rangle\\ &= \mathbb{E}[\langle Y, ic\rangle + \langle Y, b\rangle + \langle iZ, b\rangle + \langle iZ, ic\rangle]\\ &= \mathbb{E}[(-i)\langle Y, c\rangle + \langle Y, b\rangle + i\langle Z, b\rangle + \underbrace{(i)(-i)}_{-i^2=1}\langle Z, c\rangle]\\ &= \mathbb{E}(\langle Y, b\rangle + \langle Z, c\rangle) + i\mathbb{E}(\langle Z, b\rangle - \langle Y, c\rangle)\\ &= \langle \mathbb{E}Y, b\rangle + \underbrace{\langle \mathbb{E}Z, c\rangle}_{\langle i\mathbb{E}Z, ic\rangle} + i\langle \mathbb{E}Z, b\rangle - \underbrace{i\langle \mathbb{E}Y, c\rangle}_{-\langle \mathbb{E}Y, ic\rangle}\\ &= \langle \mathbb{E}Y + i\mathbb{E}Z, b\rangle + \langle \mathbb{E}Y + i\mathbb{E}Z, ic\rangle\\ &= \langle \mathbb{E}Y + i\mathbb{E}Z, a\rangle.\end{aligned}$$

♦

**Exercise 3.3**

*Solution* First, let us verify the claim in question when $\mathbf{X}$ and $\mathbf{Y}$ map to the complex plane; we will represent them by $X_1 + iX_2$ and $Y_1 + iY_2$, respectively. Based on Exercise 3.2,

$$\begin{aligned}\mathbb{E}(\mathbf{X} + \mathbf{Y}) &= \mathbb{E}(X_1 + iX_2 + Y_1 + iY_2)\\ &= \mathbb{E}[(X_1 + Y_1) + i(X_2 + Y_2)]\\ &= \mathbb{E}(X_1 + Y_1) + i\mathbb{E}(X_2 + Y_2)\\ &= \mathbb{E}X_1 + \mathbb{E}Y_1 + i\mathbb{E}X_2 + i\mathbb{E}Y_2\end{aligned}$$

$$= \underbrace{(\mathbb{E}X_1 + i\mathbb{E}X_2)}_{\mathbb{E}\mathbf{X}} + \underbrace{(\mathbb{E}Y_1 + i\mathbb{E}Y_2)}_{\mathbb{E}\mathbf{Y}}.$$

Using this result, we can prove the general case.

$$\begin{aligned}
\mathbb{E}\langle \mathbf{X} + \mathbf{Y}, \mathbf{v}\rangle &= \mathbb{E}[\langle \mathbf{X}, \mathbf{v}\rangle + \langle \mathbf{Y}, \mathbf{v}\rangle] \\
&= \mathbb{E}\langle \mathbf{X}, \mathbf{v}\rangle + \mathbb{E}\langle \mathbf{Y}, \mathbf{v}\rangle \\
&= \langle \mathbf{x}, \mathbf{v}\rangle + \langle \mathbf{y}, \mathbf{v}\rangle \\
&= \langle \mathbf{x} + \mathbf{y}, \mathbf{v}\rangle.
\end{aligned}$$

♦

## Exercise 3.4

*Solution* If $\mathbf{X}$ is a random vector taking values in the complex plane, then inner product is the ordinary product, so property (ii) of our definition of expectation says that $\mathbb{E}a\mathbf{X} = a\mathbb{E}\mathbf{X}$.

Now we will use that fact to establish the more general result for random vectors.

$$\begin{aligned}
\mathbb{E}\langle a\mathbf{X}, \mathbf{v}\rangle &= \mathbb{E}a\langle \mathbf{X}, \mathbf{v}\rangle \\
&= a\mathbb{E}\langle \mathbf{X}, \mathbf{v}\rangle \\
&= a\langle \mathbb{E}\mathbf{X}, \mathbf{v}\rangle \\
&= \langle a\mathbb{E}\mathbf{X}, \mathbf{v}\rangle.
\end{aligned}$$

♦

## Exercise 3.5

*Solution*

$$\begin{aligned}
\mathbb{E}\langle \mathbf{X}, \mathbf{v}\rangle &= \mathbb{E}\langle \mathbf{w}, \mathbf{v}\rangle \\
&= \langle \mathbf{w}, \mathbf{v}\rangle \underbrace{\mathbb{E}\mathbb{1}_\Omega}_{1} \\
&= \langle \mathbf{w}, \mathbf{v}\rangle,
\end{aligned}$$

where $\Omega$ represents the whole sample space and therefore has probability 1. ♦

## Exercise 3.12

*Solution*

$$\begin{aligned}
&\mathbb{E}(X - \mathbb{E}X)[a_1Y_1 + \ldots + a_nY_n - \mathbb{E}(a_1Y_1 + \ldots + a_nY_n)] \\
&\quad = \mathbb{E}(X - \mathbb{E}X)[a_1Y_1 + \ldots + a_nY_n - (a_1\mathbb{E}Y_1 + \ldots + a_n\mathbb{E}Y_n)]
\end{aligned}$$

$$= a_1 \underbrace{\mathbb{E}(X - \mathbb{E}X)(Y_1 - \mathbb{E}Y_1)}_{0} + \ldots + a_n \underbrace{\mathbb{E}(X - \mathbb{E}X)(Y_n - \mathbb{E}Y_n)}_{0}$$

$$= 0.$$

♦

### Exercise 3.13

*Solution* Each event has probability $1 - q$ of not occurring. The probability of an intersection of independent events equals the product of their probabilities, so the probability that *none* of the events occur is $(1 - q)^m$. The probability that *at least one* occurs is $1 - (1 - q)^m$ since it is the complement of the event that none of them occurs. ♦

### Exercise 3.14

*Solution* We can express the first inner product as $\langle \mathbf{X}, \mathbf{v} \rangle = Y + iZ$ for some random variables $Y$ and $Z$. If its expectation $\mathbb{E}Y + i\mathbb{E}Z$ is real, then $\mathbb{E}Z$ must be 0. The other inner product is the complex conjugate $\langle \mathbf{v}, \mathbf{X} \rangle = Y - iZ$. Its expectation $\mathbb{E}Y - i\mathbb{E}Z$ simplifies to $\mathbb{E}Y$ as well. ♦

### Exercise 3.20

*Solution* The $(i, j)$ entry of $\mathbb{E}\mathbb{M}\mathbb{X}$ is $\mathbb{E}\mathbf{m}_i^T \mathbf{X}_j = \mathbf{m}_i^T \mathbb{E}\mathbf{X}_j$ where $\mathbf{m}_i$ represents the $i$th row of $\mathbb{M}$ and $\mathbf{X}_j$ represents the $j$th column of $\mathbb{X}$. This is also the $(i, j)$ entry of $\mathbb{M}\mathbb{E}\mathbb{X}$. Similarly, the $(i, j)$ entry of $\mathbb{E}\mathbb{X}\mathbb{M}$ is $\mathbb{E}\mathbf{X}_i^T \mathbf{m}_j = (\mathbb{E}\mathbf{X}_i)^T \mathbf{m}_j$ where $\mathbf{X}_i$ represents the $i$th row of $\mathbb{X}$ and $\mathbf{m}_j$ represents the $j$th column of $\mathbb{M}$. ♦

### Exercise 3.21

*Solution* We will work out the matrix resulting from the multiplication then move the expectation into the matrix entries.

$$\mathbb{E}[(\mathbf{Y}-\mathbb{E}\mathbf{Y})(\mathbf{Y}-\mathbb{E}\mathbf{Y})^T]=\mathbb{E}\begin{bmatrix} Y_1-\mathbb{E}Y_1 \\ \vdots \\ Y_n-\mathbb{E}Y_n \end{bmatrix}\begin{bmatrix} Y_1-\mathbb{E}Y_1 & \cdots & Y_n-\mathbb{E}Y_n \end{bmatrix}$$

$$= \mathbb{E}\begin{bmatrix} (Y_1-\mathbb{E}Y_1)(Y_1-\mathbb{E}Y_1) & \cdots & (Y_1-\mathbb{E}Y_1)(Y_n-\mathbb{E}Y_n) \\ \vdots & \ddots & \vdots \\ (Y_n-\mathbb{E}Y_n)(Y_1-\mathbb{E}Y_1) & \cdots & (Y_n-\mathbb{E}Y_n)(Y_n-\mathbb{E}Y_n) \end{bmatrix}$$

$$= \begin{bmatrix} \mathbb{E}[(Y_1-\mathbb{E}Y_1)(Y_1-\mathbb{E}Y_1)] & \cdots & \mathbb{E}[(Y_1-\mathbb{E}Y_1)(Y_n-\mathbb{E}Y_n)] \\ \vdots & \ddots & \vdots \\ \mathbb{E}[(Y_n-\mathbb{E}Y_n)(Y_1-\mathbb{E}Y_1)] & \cdots & \mathbb{E}[(Y_n-\mathbb{E}Y_n)(Y_n-\mathbb{E}Y_n)] \end{bmatrix}$$

The empirical covariance matrix expression in Eq. 1.7 can be understood as an empirical version of this. ♦

### Exercise 3.26

*Solution* Let $\mathbf{X} := (X_1, \ldots, X_n)$, and let $\sigma_1^2, \ldots, \sigma_n^2$ represent the variances. Using Exercise 3.23,

$$\begin{aligned}
\operatorname{var} \sum_i X_i &= \operatorname{var} \mathbf{1}^T \mathbf{X} \\
&= \mathbf{1}^T \begin{bmatrix} \sigma_1^2 & & \\ & \ddots & \\ & & \sigma_n^2 \end{bmatrix} \mathbf{1} \\
&= \mathbf{1}^T \begin{bmatrix} \sigma_1^2 \\ \vdots \\ \sigma_n^2 \end{bmatrix} \\
&= \sum_i \sigma_i^2.
\end{aligned}$$

◆

### Exercise 3.36

*Solution* Let $\mathbf{Y} := (Y_1, \ldots, Y_n)$ and $\mathbf{X} := (X_1, \ldots, X_n)$. The matrix

$$(\mathbf{X} - \mu_X \mathbf{1})(\mathbf{Y} - \mu_Y \mathbf{1})^T = \begin{bmatrix} (X_1 - \mu_X)(Y_1 - \mu_Y) & \cdots & (X_1 - \mu_X)(Y_n - \mu_Y) \\ \vdots & \ddots & \vdots \\ (X_n - \mu_X)(Y_1 - \mu_Y) & \cdots & (X_n - \mu_X)(Y_n - \mu_Y) \end{bmatrix}$$

has expectation $\sigma_{X,Y}\mathbb{I}$.[1]

Let $\mathbb{J}$ be the orthogonal projection matrix onto the span of $\{\mathbf{1}\}$. We will use the same trace cyclic permutation trick that was advantageous for evaluating expected quadratic forms.

$$\begin{aligned}
\mathbb{E} \sum_i (X_i - \bar{X})(Y_i - \bar{Y}) &= \mathbb{E}(\mathbf{Y} - \bar{Y}\mathbf{1})^T (\mathbf{X} - \bar{X}\mathbf{1}) \\
&= \mathbb{E}[(\mathbb{I} - \mathbb{J})\mathbf{Y}]^T [(\mathbb{I} - \mathbb{J})\mathbf{X}] \\
&= \mathbb{E}[(\mathbb{I} - \mathbb{J})(\mathbf{Y} - \mu_Y \mathbf{1})]^T [(\mathbb{I} - \mathbb{J})(\mathbf{X} - \mu_X \mathbf{1})] \\
&= \mathbb{E}(\mathbf{Y} - \mu_Y \mathbf{1})^T (\mathbb{I} - \mathbb{J})(\mathbf{X} - \mu_X \mathbf{1}) \\
&= \mathbb{E} \operatorname{tr} [(\mathbf{Y} - \mu_Y \mathbf{1})^T (\mathbb{I} - \mathbb{J})(\mathbf{X} - \mu_X \mathbf{1})]
\end{aligned}$$

[1] In general, $\mathbb{E}(\mathbf{X} - \mathbb{E}\mathbf{X})(\mathbf{Y} - \mathbb{E}\mathbf{Y})^T$ is often called the "cross-covariance matrix" of $\mathbf{X}$ and $\mathbf{Y}$.

$$= \operatorname{tr}[(\mathbb{I} - \mathbb{J})\underbrace{\mathbb{E}(\mathbf{X} - \boldsymbol{\mu}_X\mathbf{1})(\mathbf{Y} - \boldsymbol{\mu}_Y\mathbf{1})^T}_{\sigma_{X,Y}\mathbb{I}}]$$

$$= (n-1)\sigma_{X,Y}.$$

♦

### Exercise 4.3

*Solution* The *least-squares line* has $\hat{\alpha} = \overline{Y}$, and its expectation is

$$\begin{aligned}
\mathbb{E}\overline{Y} &= \mathbb{E}(\tfrac{1}{n}\sum_i Y_i) \\
&= \tfrac{1}{n}\sum_i \underbrace{\mathbb{E}Y_i}_{\alpha+\beta(x_i-\bar{x})} \\
&= \alpha + \beta\tfrac{1}{n}\underbrace{\sum_i (x_i - \bar{x})}_{0} \\
&= \alpha.
\end{aligned}$$

The other coefficient's expectation is

$$\begin{aligned}
\mathbb{E}\hat{\beta} &= \mathbb{E}\frac{\langle \mathbf{x} - \bar{x}\mathbf{1}, \mathbf{Y} - \overline{Y}\mathbf{1}\rangle}{\|\mathbf{x} - \bar{x}\mathbf{1}\|^2} \\
&= \frac{\langle \mathbf{x} - \bar{x}\mathbf{1}, \mathbb{E}\mathbf{Y} - \mathbb{E}\overline{Y}\mathbf{1}\rangle}{\|\mathbf{x} - \bar{x}\mathbf{1}\|^2} \\
&= \frac{\langle \mathbf{x} - \bar{x}\mathbf{1}, [\alpha\mathbf{1} + \beta(\mathbf{x} - \bar{x}\mathbf{1})] - \alpha\mathbf{1}\rangle}{\|\mathbf{x} - \bar{x}\mathbf{1}\|^2} \\
&= \beta\frac{\langle \mathbf{x} - \bar{x}\mathbf{1}, \mathbf{x} - \bar{x}\mathbf{1}\rangle}{\|\mathbf{x} - \bar{x}\mathbf{1}\|^2} \\
&= \beta.
\end{aligned}$$

♦

### Exercise 4.4

*Solution* The variance of $\hat{a}$ works out to be $\frac{\sigma^2}{n}$, exactly as in Exercise 4.2. The variance of $\hat{b}$ is

$$\operatorname{var}\hat{b} = \operatorname{var}\frac{\frac{1}{n}\langle \mathbf{x} - \bar{x}\mathbf{1}, \mathbf{Y}\rangle}{\sigma_{\mathbf{x}}^2}$$

$$
\begin{aligned}
&= \frac{1}{n^2\sigma_{\mathbf{x}}^4} \operatorname{cov}(\mathbf{x} - \bar{x}\mathbf{1})^T \mathbf{Y} \\
&= \frac{1}{n^2\sigma_{\mathbf{x}}^4} (\mathbf{x} - \bar{x}\mathbf{1})^T \sigma^2 \mathbb{I} (\mathbf{x} - \bar{x}\mathbf{1}) \\
&= \frac{\sigma^2}{n^2\sigma_{\mathbf{x}}^4} \underbrace{\|\mathbf{x} - \bar{x}\mathbf{1}\|^2}_{n\sigma_{\mathbf{x}}^2} \\
&= \frac{\sigma^2}{n\sigma_{\mathbf{x}}^2}
\end{aligned}
$$

unless $\sigma_{\mathbf{x}}^2 = 0$ in which case $\hat{\beta} \equiv 0$ has variance 0. These two variances are the diagonals of the covariance matrix. The off-diagonals are equal to the covariance between $\hat{a}$ and $\hat{b}$. It will be important to realize that the average of the entries of $\mathbb{E}\mathbf{Y}$ is

$$
\begin{aligned}
\tfrac{1}{n}\mathbf{1}^T \mathbb{E}\mathbf{Y} &= \tfrac{1}{n}\mathbf{1}^T [\alpha \mathbf{1} - \beta(\mathbf{x} - \bar{x}\mathbf{1})] \\
&= \alpha - \beta(\bar{x} - \bar{x}) \\
&= \alpha.
\end{aligned}
$$

Thus $\overline{Y} - \alpha$ can be rewritten as $\frac{1}{n}\mathbf{1}^T(\mathbf{Y} - \mathbb{E}\mathbf{Y})$.

$$
\begin{aligned}
\mathbb{E}(\overline{Y} - \alpha) \left( \frac{\frac{1}{n}\langle \mathbf{x} - \bar{x}\mathbf{1}, \mathbf{Y} \rangle}{\sigma_{\mathbf{x}}^2} \right) &= \mathbb{E} \frac{1}{n^2\sigma_{\mathbf{x}}^2} \mathbf{1}^T (\mathbf{Y} - \mathbb{E}\mathbf{Y}) \mathbf{Y}^T (\mathbf{x} - \bar{x}\mathbf{1}) \\
&= \frac{1}{n^2\sigma_{\mathbf{x}}^2} \mathbf{1}^T [\underbrace{\mathbb{E}(\mathbf{Y} - \mathbb{E}\mathbf{Y})\mathbf{Y}^T}_{\sigma^2\mathbb{I}}] (\mathbf{x} - \bar{x}\mathbf{1}) \\
&= \frac{\sigma^2}{n\sigma_{\mathbf{x}}^2} \underbrace{\frac{1}{n}\mathbf{1}^T (\mathbf{x} - \bar{x}\mathbf{1})}_{\bar{x} - \bar{x}} \\
&= 0.
\end{aligned}
$$

◆

### Exercise 4.9

*Solution* Let $\widetilde{\mathbb{X}}$ be the centered version of the explanatory data matrix, and let $\Sigma_j^-$ be the $j$th row of the generalized inverse of its empirical covariance matrix (as a column vector). Borrowing tricks from Exercise 4.4, the covariance between $\hat{\alpha}$ and $\hat{\beta}_j$ is

$$
\mathbb{E}(\overline{Y} - \alpha)((\Sigma_j^-)^T \tfrac{1}{n} \widetilde{\mathbb{X}}^T \mathbf{Y}) = \mathbb{E} \tfrac{1}{n^2} \mathbf{1}^T (\mathbf{Y} - \mathbb{E}\mathbf{Y}) \mathbf{Y}^T \widetilde{\mathbb{X}} \Sigma_j^-
$$

$$= \frac{1}{n^2}\mathbf{1}^T[\underbrace{\mathbb{E}(\mathbf{Y}-\mathbb{E}\mathbf{Y})\mathbf{Y}^T}_{\text{cov}\,\mathbf{Y}=\sigma^2\mathbb{I}}]\widetilde{\mathbb{X}}\Sigma_j^-$$

$$= \frac{\sigma^2}{n}\underbrace{\frac{1}{n}\mathbf{1}^T\widetilde{\mathbb{X}}}_{\mathbf{0}^T}\Sigma_j^-$$

$$= 0.$$

♦

## Exercise 4.12

*Solution* First, consider the claim of the Gauss–Markov theorem: the variance of $\hat{\boldsymbol{\gamma}}^T\mathbf{v}$ is no greater than the variance of $\check{\boldsymbol{\gamma}}^T\mathbf{v}$. An alternative expression for the squared deviation of $\hat{\boldsymbol{\gamma}}^T\mathbf{v}$ from its mean is

$$\begin{aligned}(\hat{\boldsymbol{\gamma}}^T\mathbf{v} - \mathbb{E}\hat{\boldsymbol{\gamma}}^T\mathbf{v})^2 &= (\hat{\boldsymbol{\gamma}}^T\mathbf{v} - \boldsymbol{\gamma}^T\mathbf{v})^2 \\ &= [(\hat{\boldsymbol{\gamma}} - \boldsymbol{\gamma})^T\mathbf{v}][(\hat{\boldsymbol{\gamma}} - \boldsymbol{\gamma})^T\mathbf{v}] \\ &= (\hat{\boldsymbol{\gamma}} - \boldsymbol{\gamma})^T\mathbf{v}\mathbf{v}^T(\hat{\boldsymbol{\gamma}} - \boldsymbol{\gamma}),\end{aligned}$$

and likewise for $\check{\boldsymbol{\gamma}}$. The *expected* squared deviation is the variance, so Gauss–Markov tells us that $\mathbb{E}(\hat{\boldsymbol{\gamma}} - \boldsymbol{\gamma})^T\mathbf{v}\mathbf{v}^T(\hat{\boldsymbol{\gamma}} - \boldsymbol{\gamma})$ is no greater than $(\check{\boldsymbol{\gamma}} - \boldsymbol{\gamma})^T\mathbf{v}\mathbf{v}^T(\check{\boldsymbol{\gamma}} - \boldsymbol{\gamma})$ for every $\mathbf{v}$.

With a spectral decomposition for $\mathbb{L}$,

$$\begin{aligned}(\hat{\boldsymbol{\gamma}} - \boldsymbol{\gamma})^T\mathbb{L}(\hat{\boldsymbol{\gamma}} - \boldsymbol{\gamma}) &= (\hat{\boldsymbol{\gamma}} - \boldsymbol{\gamma})^T(\lambda_1\mathbf{q}_1\mathbf{q}_1^T + \ldots + \lambda_d\mathbf{q}_d\mathbf{q}_d^T)(\hat{\boldsymbol{\gamma}} - \boldsymbol{\gamma}) \\ &= \lambda_1(\hat{\boldsymbol{\gamma}} - \boldsymbol{\gamma})^T\mathbf{q}_1\mathbf{q}_1^T(\hat{\boldsymbol{\gamma}} - \boldsymbol{\gamma}) + \ldots + \lambda_d(\hat{\boldsymbol{\gamma}} - \boldsymbol{\gamma})^T\mathbf{q}_d\mathbf{q}_d^T)(\hat{\boldsymbol{\gamma}} - \boldsymbol{\gamma}).\end{aligned}$$

Each eigenvalue is non-negative, so each term is no greater than the corresponding expression with $\check{\boldsymbol{\gamma}}$ in place of $\hat{\boldsymbol{\gamma}}$. ♦

## Exercise 4.13

*Solution* Let $\mathbf{v}_{n+1}^T = \mathbf{w}^T\mathbb{M}$. The expectation of the new response value can be represented as the $\mathbf{w}$ linear combination of the expectations of previous response values:

$$\begin{aligned}\mathbb{E}Y_{n+1} &= \mathbf{v}_{n+1}^T\boldsymbol{\gamma} \\ &= \mathbf{w}^T\mathbb{M}\boldsymbol{\gamma} \\ &= \mathbf{w}^T\mathbb{E}\mathbf{Y}.\end{aligned}$$

The predictor can be written as the same linear combination of the previous predicted values:

$$\begin{aligned}\widehat{Y}_{n+1} &= \mathbf{v}_{n+1}^T \hat{\boldsymbol{y}} \\ &= \mathbf{w}^T \mathbb{M} \hat{\boldsymbol{y}} \\ &= \mathbf{w}^T \widehat{\mathbf{Y}}.\end{aligned}$$

Because $\widehat{\mathbf{Y}}$ is unbiased for $\mathbb{E}\mathbf{Y}$, we have

$$\begin{aligned}\mathbb{E}\widehat{Y}_{n+1} &= \mathbf{w}^T \mathbb{E}\widehat{\mathbf{Y}} \\ &= \mathbf{w}^T \mathbb{E}\mathbf{Y}\end{aligned}$$

which is the same expectation we found for $\mathbb{E}Y_{n+1}$. ♦

**Exercise 4.14**

*Solution* The expectation of the response variable is

$$\begin{aligned}\mathbb{E}\mathbf{Y} &= b_1\mathbf{x} + b_2\mathbf{z} \\ &= b_1 a\mathbf{z} + b_2\mathbf{z} \\ &= (ab_1 + b_2)\mathbf{z} \\ &= c\mathbf{z}.\end{aligned}$$

Every possible value of $a$ implies a different expectation for the response variable, so it can be estimated. In fact the orthogonal projection's coefficient $\frac{\langle \mathbf{Y}, \mathbf{z}\rangle}{\|\mathbf{z}\|^2}$ is unbiased based on Exercise 4.10. ♦

**Exercise 4.16**

*Solution*

$$\begin{aligned}\mathbb{E}(Y_{n+1} - \widehat{Y}_{n+1})^2 &= \mathbb{E}[(Y_{n+1} - \mathbb{E}Y_{n+1}) - (\widehat{Y}_{n+1} - \mathbb{E}Y_{n+1})]^2 \\ &= \mathbb{E}(Y_{n+1} - \mathbb{E}Y_{n+1})^2 - 2\mathbb{E}(Y_{n+1} - \mathbb{E}Y_{n+1})(\widehat{Y}_{n+1} - \mathbb{E}Y_{n+1}) \\ &\quad + \mathbb{E}(\widehat{Y}_{n+1} - \mathbb{E}Y_{n+1})^2 \\ &= \operatorname{var} Y_{n+1} - 2\underbrace{\mathbb{E}(Y_{n+1} - \mathbb{E}Y_{n+1})}_{0}\mathbb{E}(\widehat{Y}_{n+1} - \mathbb{E}Y_{n+1}) \\ &\quad + \mathbb{E}(\widehat{Y}_{n+1} - \mathbb{E}Y_{n+1})^2 = \operatorname{var} Y_{n+1} + \mathbb{E}(\widehat{Y}_{n+1} - \mathbb{E}Y_{n+1})^2.\end{aligned}$$

♦

### Exercise 5.2

*Solution* From Exercise 3.23, we calculate the covariances to be $\mathbb{M}_1\mathbb{M}_1^T$ and $\mathbb{M}_2\mathbb{M}_2^T$.

Suppose $\mathbb{M}_1$ has the singular value decomposition $\mathbb{U}\mathbb{S}\mathbb{V}_1^T$. Then the spectral decomposition of $\mathbb{M}_1\mathbb{M}_1^T$ is $\mathbb{U}\mathbb{S}\mathbb{U}^T$. By the assumption that the covariances are equal, we see that $\mathbb{M}_2\mathbb{M}_2^T$ must also be equal to $\mathbb{U}\mathbb{S}\mathbb{U}^T$. Thus, a singular value decomposition of $\mathbb{M}_2$ has the same matrix $\mathbb{U}$ on the left and the same matrix of singular values; we will write $\mathbb{M}_2 = \mathbb{U}\mathbb{S}\mathbb{V}_2^T$.

We need to compare the distributions of $\mathbb{U}\mathbb{S}\mathbb{V}_1^T\mathbf{Z}_1$ and $\mathbb{U}\mathbb{S}\mathbb{V}_2^T\mathbf{Z}_2$. The entries of $\mathbb{V}_1^T\mathbf{Z}_1$ are the coordinates of $\mathbf{Z}_1$ with respect to the orthonormal columns of $\mathbb{V}$, so they are iid standard Normal according to our discussion in Sect. 5.1. Likewise, the entries of $\mathbb{V}_2^T\mathbf{Z}_2$ are standard Normal, so we can conclude that the two random vectors in question have the same distribution. ♦

### Exercise 5.4

*Solution* Let $\mathbf{X}_1 \sim N(\boldsymbol{\mu}_1, \mathbb{C}_1)$ and $\mathbf{X}_2 \sim N(\boldsymbol{\mu}_2, \mathbb{C}_2)$. With standard Normal $\mathbf{Z}_1$ and $\mathbf{Z}_2$, we can represent the sum as

$$\begin{aligned}\mathbf{X}_1 + \mathbf{X}_2 &= (\mathbb{C}_1^{1/2}\mathbf{Z}_1 + \boldsymbol{\mu}_1) + (\mathbb{C}_2^{1/2}\mathbf{Z}_2 + \boldsymbol{\mu}_2) \\ &= \begin{bmatrix}\mathbb{C}_1^{1/2} & \mathbb{C}_2^{1/2}\end{bmatrix}\begin{bmatrix}\mathbf{Z}_1 \\ \mathbf{Z}_2\end{bmatrix} + [\boldsymbol{\mu}_1 + \boldsymbol{\mu}_2].\end{aligned}$$

We are almost finished, but consider carefully the vector $\begin{bmatrix}\mathbf{Z}_1 \\ \mathbf{Z}_2\end{bmatrix}$ that has the entries of $\mathbf{Z}_1$ stacked on top of the entries of $\mathbf{Z}_2$. We cannot assume that the entries of $\mathbf{Z}_1$ are independent of the entries of $\mathbf{Z}_2$, so the stacked vector is not necessarily standard Normal. However, with $\mathbb{C}$ representing the covariance matrix of $\begin{bmatrix}\mathbf{Z}_1 \\ \mathbf{Z}_2\end{bmatrix}$ and with $\mathbf{Z}$ a standard Normal random vector of the same size as $\begin{bmatrix}\mathbf{Z}_1 \\ \mathbf{Z}_2\end{bmatrix}$, then we can rewrite the expression as

$$\begin{bmatrix}\mathbb{C}_1^{1/2} & \mathbb{C}_2^{1/2}\end{bmatrix}\mathbb{C}^{1/2}\mathbf{Z} + [\boldsymbol{\mu}_1 + \boldsymbol{\mu}_2]$$

which fits the definition of a Normal random vector. ♦

### Exercise 5.5

*Solution* Let $Z \sim N(0, 1)$. Independently of $Z$, let $B$ take values $-1$ and 1 each with probability. Finally, define $Y := BZ$. By inspecting the cdf of $Y$,

$$\mathbb{P}(Y \leq t) = \mathbb{P}(B = 1 \cap Z \leq t) + \mathbb{P}(B = -1 \cap Z \geq t)$$

$$
\begin{aligned}
&= \mathbb{P}(B = 1)\mathbb{P}(Z \le t) + \mathbb{P}(B = -1)\mathbb{P}(Z \ge -t) \\
&= \mathbb{P}(B = 1)\mathbb{P}(Z \le t) + \mathbb{P}(B = -1)\underbrace{\mathbb{P}(Z \ge -t)}_{\mathbb{P}(Z \le t)} \\
&= \underbrace{[\mathbb{P}(B = 1) + \mathbb{P}(B = -1)]}_{1}\,\mathbb{P}(Z \le t)
\end{aligned}
$$

we see that it is also standard Normal as it has the same cdf as $Z$. If you learn that $Z = z$, you know that $Y$ is either $z$ or $-z$, so $Z$ and $Y$ clearly are not independent. However, their correlation is

$$
\begin{aligned}
\mathbb{E}ZY &= \mathbb{E}Z(BZ) \\
&= (\underbrace{\mathbb{E}B}_{0})(\mathbb{E}Z^2) \\
&= 0.
\end{aligned}
$$

♦

# Bibliography


M. Artin, *Algebra* (Pearson, New York, 2010)

W.D. Brinda, *Adaptive Estimation with Gaussian Radial Basis Mixtures*. Ph.D. thesis, Yale University, 2018

P.C. Brown, H.L. Roedinger, M.A. McDaniel, *Make It Stick: The Science of Successful Learning* (Belknap Press, Cambridge, 2014)

N. Carr, *The Shallows: What the Internet is Doing to our Brains* (W. W. Norton and Company, New York, 2010)

R. Christensen, *Plane Answers to Complex Questions: The Theory of Linear Models* (Springer, New York, 2011)

K.A. Ericsson, R. Pool, *Peak: Secrets from the New Science of Expertise* (Eamon Dolan/Mariner Books, Boston, 2017)

J.E. Gentle, *Matrix Algebra: Theory, Computations and Applications in Statistics* (Springer, Cham, 2017)

D. Goleman, *Focus: The Hidden Driver of Excellence* (HarperCollins Publishers, New York, 2013)

M. Grötschel, L. Lovász, A. Schrijver, *Geometric Algorithms and Combinatorial Optimization* (Springer, New York, 1988)

C. Newport, *Deep Work: Rules for Focused Success in a Distracted World* (Grand Central Publishing, New York, 2016)

R.C. Penney, *Linear Algebra: Ideas and Applications* (Wiley, New York, 2015)

D. Pollard, *A User's Guide to Measure Theoretic Probability* (Cambridge University Press, New York, 2002)

E. Schechter, *Handbook of Analysis and its Foundations* (Academic Press, San Diego, 1997)

T.M. Sterner, *The Practicing Mind: Developing Focus and Discipline in Your Life* (New World Library, Novata, 2012)

G.A. Young, R.L. Smith, *Essentials of Statistical Inference* (Cambridge University Press, New York, 2005)




W. D. Brinda, *Visualizing Linear Models*,
https://doi.org/10.1007/978-3-030-64167-2

# Index




W. D. Brinda, *Visualizing Linear Models*,
https://doi.org/10.1007/978-3-030-64167-2

Printed in the United States
by Baker & Taylor Publisher Services